Introductory Chemical Engineering

Introductory Chemical Engineering

Introductory Chemical Engineering

Edited by
Connor Wilson

Larsen & Keller
www.larsen-keller.com

Introductory Chemical Engineering
Edited by Connor Wilson
ISBN: 978-1-63549-063-3 (Hardback)

© 2017 Larsen & Keller

▤ Larsen & Keller

Published by Larsen and Keller Education,
5 Penn Plaza,
19th Floor,
New York, NY 10001, USA

Cataloging-in-Publication Data

Introductory chemical engineering / edited by Connor Wilson.
    p. cm.
Includes bibliographical references and index.
ISBN 978-1-63549-063-3
1. Chemical engineering. 2. Chemistry, Technical. I. Wilson, Connor.
TP155 .I58 2017
660--dc23

The publisher's policy is to use permanent paper from mills that operate a sustainable forestry policy. Furthermore, the publisher ensures that the text paper and cover boards used have met acceptable environmental accreditation standards.

Printed and bound in the United States of America.

For more information regarding Larsen and Keller Education and its products, please visit the publisher's website www.larsen-keller.com

Table of Contents

Preface   **VII**

Chapter 1 **Introduction to Chemical Engineering**   **1**
           a. Chemical Engineering   1
           b. Raw Material   6
           c. Chemical Substance   7
           d. Chemical Element   14

Chapter 2 **Branches of Chemical Engineering**   **39**
           a. Biochemical Engineering   39
           b. Electrochemical Engineering   39

Chapter 3 **Processes in Chemical Engineering**   **41**
           a. Chemical Process   41
           b. Chemical Reaction   59
           c. Unit Operation   117
           d. Process Design   119
           e. Chemical Process Modeling   121
           f. Process Integration   121

Chapter 4 **Separation Process: Types and Techniques**   **124**
           a. Separation Process   124
           b. Chromatography   127
           c. Crystallization   140
           d. Distillation   151
           e. Membrane Technology   171
           f. Field Flow Fractionation   179
           g. Theoretical Plate   184

Chapter 5 **Major Aspects of Chemical Engineering**   **188**
           a. Transport Phenomena   188
           b. Stefan Tube   193
           c. Chemical Reactor   194
           d. Chemical Plant   199
           e. Chemical Thermodynamics   208
           f. Process Simulation   215
           g. Chemical Kinetics   218
           h. Process Flow Diagram   223
           i. Process Miniaturization   226

Chapter 6 **Applications of Chemical Engineering**   **232**
           a. Food Engineering   232

b. Plastics Engineering   234
c. Pharmaceutical Engineering   235
d. Tissue Engineering   237
e. Ceramic Engineering   251
f. Chemometrics   268
g. Process Engineering   273

Chapter 7 **Softwares Related to Chemical Engineering**   **276**
a. Chemical WorkBench   276
b. COCO Simulator   277
c. DWSIM   278
d. ProMax   279

Chapter 8 **Evolution of Chemical Engineering**   **281**

Permissions

Index

Preface

This book attempts to understand the multiple branches that fall under the discipline of chemical engineering and how such concepts have practical applications. Chemical engineering is an intricate subject which includes other fields like applied mathematics, life sciences, microbiology, economics and biochemistry in order to transport, produce, transform chemicals, energy and materials. Most of the topics introduced in this book cover new techniques and the applications of chemical engineering. Such selected concepts that redefine this field have been presented in this text. Also included are the industrial applications related to this field. Different approaches, evaluations and methodologies have been included in it. The text will serve as a valuable source of reference for those interested in chemical engineering.

A detailed account of the significant topics covered in this book is provided below:

Chapter 1- Chemical engineering is a branch of engineering that applies life sciences with applied mathematics and economics. Raw materials, chemical substances and chemical elements are some of the themes explained in the following text. This chapter is an overview of the subject matter incorporating all the major aspects of chemical engineering.

Chapter 2- Chemical engineering is an interdisciplinary subject. It has numerous branches; some of these are biochemical engineering, electrochemical engineering and chemical process modeling. Biochemical engineering mainly deals with designing and construction of unit processes whereas electrochemical engineering deals with the functions of electrochemical phenomena. This text will provide a glimpse of the related fields of chemical engineering briefly.

Chapter 3- Chemical reactions are processes that transform one set of chemical substances into another. Chemical processes are processes that change chemicals or chemical compounds. Other themes included in this section include unit operation, process design, chemical process modeling and process integration. This chapter discusses the methods of chemical engineering in a critical manner providing key analysis to the subject matter.

Chapter 4- Separation processes are methods that are used to convert mixtures of chemical substances into two or more distinct mixtures. The technical term used for the separation of mixtures is known as chromatography. The techniques elucidated in this section are of vital importance, and provides a better understanding of separation processes.

Chapter 5- Transport phenomena, Stefan tube, chemical reactor, process simulation and chemical kinetics are some of the major aspects of chemical engineering. Transport phenomena deals with the exchange of mass and energy between systems whereas Stefan tubes are devices used for measuring diffusion coefficients. Chemical engineering is best understood in confluence with the major topics listed in the following text.

Chapter 6- Chemical engineering has diverse applications. Some of these are food engineering, plastics engineering, tissue engineering, ceramic engineering, cheminformatics and process engineering. Plastics engineering deals with the designing and manufacturing of plastic products and food engineering deals includes agricultural engineering, mechanical and chemical engineering. The diverse applications of chemical engineering in the current scenario have been thoroughly discussed in this chapter.

Chapter 7- Chemical workbench is a software that aims at the scale of homogeneous gas- phase and heterogeneous processes whereas COCO simulator is a free-of- charge and sequential simulation process modeling environment. In order to completely understand chemical engineering, it is necessary to understand the softwares related to it. The following text elucidates the varied softwares used in chemical engineering.

Chapter 8- Chemical engineering has evolved over the years. The Industrial Revolution was a major landmark in the mass manufacture of chemicals. Chemical engineering in contemporary times is used in numerous fields, and is studied in a number of universities. This chapter has been carefully written to provide an easy understanding of the development of chemical engineering.

It gives me an immense pleasure to thank our entire team for their efforts. Finally in the end, I would like to thank my family and colleagues who have been a great source

Editor

Introduction to Chemical Engineering

Chemical engineering is a branch of engineering that applies life sciences with applied mathematics and economics. Raw materials, chemical substances and chemical elements are some of the themes explained in the following text. This chapter is an overview of the subject matter incorporating all the major aspects of chemical engineering.

Chemical Engineering

Chemical engineering is a branch of engineering that applies physical sciences (physics and chemistry) and life sciences (microbiology) and together with applied mathematics and economics to produce, transform, transport, and properly use chemicals, materials and energy. Essentially, chemical engineers design large-scale processes that convert chemicals, raw materials, living cells, microorganisms and energy into useful forms and products.

Chemical engineers design, construct and operate process plants (distillation columns pictured)

Etymology

A 1996 *British Journal for the History of Science* article cites James F. Donnelly for mentioning an 1839 reference to chemical engineering in relation to the production of sulfuric acid. In the same paper however, George E. Davis, an English consultant, was credited for having coined the term. The *History of Science in United States: An Encyclopedia* puts this at around 1890. "Chemical engineering", describing the use of mechanical equipment in the chemical industry, became common vocabulary in England after 1850. By 1910, the profession, "chemical engineer," was already in common use in Britain and the United States.

George E. Davis

History

Chemical engineering emerged upon the development of unit operations, a fundamental concept of the discipline of chemical engineering. Most authors agree that Davis invented unit operations if not substantially developed it. He gave a series of lectures on unit operations at the Manchester Technical School (later part of the University of Manchester) in 1887, considered to be one of the earliest such about chemical engineering. Three years before Davis' lectures, Henry Edward Armstrong taught a degree course in chemical engineering at the City and Guilds of London Institute. Armstrong's course "failed simply because its graduates ... were not especially attractive to employers." Employers of the time would have rather hired chemists and mechanical engineers. Courses in chemical engineering offered by Massachusetts Institute of Technology (MIT) in the United States, Owens College in Manchester, England, and University College London suffered under similar circumstances.

Students inside an industrial chemistry laboratory at MIT

Starting from 1888, Lewis M. Norton taught at MIT the first chemical engineering course in the United States. Norton's course was contemporaneous and essentially similar with Armstrong's course. Both courses, however, simply merged chemistry and engineering subjects. "Its practitioners had difficulty convincing engineers that they were engineers and chemists that they were not simply chemists." Unit operations was introduced into the course by William Hultz Walker in 1905. By the early 1920s, unit operations became an important aspect of chemical engineering at MIT and other US universities, as well as at Imperial College London. The American Institute of Chemical Engineers (AIChE), established in 1908, played a key role in making chemical engineering considered an independent science, and unit operations central to chemical engineering. For instance, it defined chemical engineering to be a "science of itself, the basis of which is ... unit operations" in a 1922 report; and with which principle, it had published a list of academic institutions which offered "satisfactory" chemical engineering courses. Meanwhile, promoting chemical engineering as a distinct science in Britain lead to the establishment of the Institution of Chemical Engineers (IChemE) in 1922. IChemE likewise helped make unit operations considered essential to the discipline.

New Concepts and Innovations

By the 1940s, it became clear that unit operations alone was insufficient in developing chemical reactors. While the predominance of unit operations in chemical engineering courses in Britain and the United States continued until the 1960s, transport phenomena started to experience greater focus. Along with other novel concepts, such process systems engineering (PSE), a "second paradigm" was defined. Transport phenomena gave an analytical approach to chemical engineering while PSE focused on its synthetic elements, such as control system and process design. Developments in chemical engineering before and after World War II were mainly incited by the petrochemical industry, however, advances in other fields were made as well. Advancements in biochemical engineering in the 1940s, for example, found application in the pharmaceutical industry, and allowed for the mass production of various antibiotics, including penicillin and streptomycin. Meanwhile, progress in polymer science in the 1950s paved way for the "age of plastics".

Safety and Hazard Developments

Concerns regarding the safety and environmental impact of large-scale chemical man-ufacturing facilities were also raised during this period. *Silent Spring*, published in 1962, alerted its readers to the harmful effects of DDT, a potent insecticide. The 1974 Flixborough disaster in the United Kingdom resulted in 28 deaths, as well as damage to a chemical plant and three nearby villages. The 1984 Bhopal disaster in India resulted in almost 4,000 deaths. These incidents, along with other incidents, affected the repu-tation of the trade as industrial safety and environmental protection were given more focus. In response, the IChemE required safety to be part of every degree course that it accredited after 1982. By the 1970s, legislation and monitoring agencies were instituted in various countries, such as France, Germany, and the United States.

Recent Progress

Advancements in computer science found applications designing and managing plants, simplifying calculations and drawings that previously had to be done manually. The completion of the Human Genome Project is also seen as a major development, not only advancing chemical engineering but genetic engineering and genomics as well. Chemical engineering principles were used to produce DNA sequences in large quan-tities.

Concepts

Chemical engineering involves the application of several principles. Key concepts are presented below.

Chemical Reaction Engineering

Chemical engineering involves managing plant processes and conditions to ensure op-timal plant operation. Chemical reaction engineers construct models for reactor anal-ysis and design using laboratory data and physical parameters, such as chemical ther-modynamics, to solve problems and predict reactor performance.

Plant Design and Construction

Chemical engineering design concerns the creation of plans, specification, and econom-ic analyses for pilot plants, new plants or plant modifications. Design engineers often work in a consulting role, designing plants to meet clients' needs. Design is limited by a number of factors, including funding, government regulations and safety standards. These constraints dictate a plant's choice of process, materials and equipment.

Plant construction is coordinated by project engineers and project managers depend-ing on the size of the investment. A chemical engineer may do the job of project engi-neer full-time or part of the time, which requires additional training and job skills, or act as a consultant to the project group. In USA the education of chemical engineering graduates from the Baccalaureate programs accredited by ABET do not usually stress project engineering education, which can be obtained by specialized training, as electives, or from graduate programs. Project engineering jobs are some of the largest employers for chemical engineers.

Process Design and Analysis

A unit operation is a physical step in an individual chemical engineering process. Unit operations (such as crystallization, filtration, drying and evaporation) are used to prepare reactants, purifying and separating its products, recycling unspent reactants, and controlling energy transfer in reactors. On the other hand, a unit process is the chemical equivalent of a unit operation. Along with unit operations, unit processes constitute a process operation. Unit processes (such as nitration and oxidation) involve the conversion of material by biochemical, thermochemical and other means. Chemical engineers responsible for these are called process engineers.

Process design requires the definition of equipment types and sizes as well as how they are connected together and the materials of construction. Details are often printed on a Process Flow Diagram which is used to control the capacity and reliability of a new or modified chemical factory.

Education for chemical engineers in the first college degree 3 or 4 years of study stresses the principles and practices of process design. The same skills are used in existing chemical plants to evaluate the efficiency and make recommendations for improvements.

Transport Phenomena

Transport phenomena occur frequently in industrial problems. These include fluid dynamics, heat transfer and mass transfer, which mainly concern momentum transfer, energy transfer and transport of chemical species respectively. Basic equations for describing the three phenomena in the macroscopic, microscopic and molecular levels are very similar. Thus, understanding transport phenomena requires a thorough understanding of mathematics.

Applications and Practice

Operators in a chemical plant using an older analog control board, seen in East-Germany, 1986.

Chemical engineers "develop economic ways of using materials and energy". Chemical engineers use chemistry and engineering to turn raw materials into usable products, such as medicine, petrochemicals and plastics on a large-scale, industrial setting. They are also involved in waste management and research. Both applied and research facets could make extensive use of computers.

Chemical engineers use computers to control automated systems in plants.

A chemical engineer may be involved in industry or university research where they are tasked in designing and performing experiments to create new and better ways of production, controlling pollution, conserving resources and making these processes safer. They may be involved in designing and constructing plants as a project engineer. In this field, the chemical engineer uses their knowledge in selecting plant equipment and the optimum method of production to minimize costs and increase profitability. After its construction, they may help in upgrading its equipment. They may also be involved in its daily operations. Chemical engineers may be permanently employed at chemical plants to manage operations. Alternatively, they may serve in a consultant role to troubleshoot problems, manage process changes and otherwise assist plant operators.

Raw Material

Anthracite coal stockpile at the shipping area of a mid-slope coal processing plant. The road leading up beyond the stockpile leads down from a Mountain top strip mine which sources the materials.

Latex being collected from a tapped rubber tree

A raw material, also known as a feedstock or most correctly unprocessed material, is a basic material that is used to produce goods, finished products, energy, or intermediate materials which are feedstock for future finished products. As feedstock, the term connotes these materials are bottleneck assets and are highly important with regards to producing other products. An example of this is crude oil, which is a raw material and a feedstock used in the production of industrial chemicals, fuels, plastics, and pharmaceutical goods; lumber is a raw material used to produce a variety of products including furniture.

The term "raw material" denotes materials in minimally processed or unprocessed in states; e.g., raw latex, crude oil, cotton, coal, raw biomass, iron ore, air, logs, or seawater i.e. "...*any product of agriculture, forestry, fishing and any other mineral that is in its natural form or which has undergone the transformation required to prepare it for internationally marketing in substantial volumes.*"

Africa has 30% of global reserves of non-energy mineral raw materials but this wealth has had a negative impact upon African countries. This phenomenon, known as "Dutch disease" or the "resource curse", occurs when the economy of a country is mainly based upon its exports due to its method of governance. An example of this is the "geological scandal" of the Democratic Republic of Congo as it is rich in raw materials; the Second Congo War focused on controlling these raw materials.

Raw materials are also used by non-humans, such as birds using found objects and twigs to create nests.

Chemical Substance

A chemical substance is a form of matter that has constant chemical composition and characteristic properties. It cannot be separated into components by physical separation methods, i.e., without breaking chemical bonds. Chemical substances can be chemical elements, chemical compounds, ions or alloys.

Steam and liquid water are two different forms of the same chemical substance, water.

Chemical substances are often called 'pure' to set them apart from mixtures. A common example of a chemical substance is pure water; it has the same properties and the same ratio of hydrogen to oxygen whether it is isolated from a river or made in a laboratory. Other chemical substances commonly encountered in pure form are diamond (carbon), gold, table salt (sodium chloride) and refined sugar (sucrose). However, in practice, no substance is entirely pure, and chemical purity is specified according to the intended use of the chemical.

Chemical substances exist as solids, liquids, gases or plasma, and may change between these phases of matter with changes in temperature or pressure. Chemical substances may be combined or converted to others by means of chemical reactions.

Forms of energy, such as light and heat, are not considered to be matter, and are thus not "substances" in this regard.

Definition

A chemical substance (also called a pure substance) may well be defined as "any material with a definite chemical composition" in an introductory general chemistry textbook. According to this definition a chemical substance can either be a pure chemical element or a pure chemical compound. But, there are exceptions to this definition; a pure substance can also be defined as a form of matter that has both definite composition and distinct properties. The chemical substance index published by CAS also includes several alloys of uncertain composition. Non-stoichiometric compounds are a special case (in inorganic chemistry) that violates the law of constant composition, and for them, it is sometimes difficult to draw the line between a mixture and a compound, as in the case of palladium hydride. Broader definitions of chemicals or chemical substances can be found, for example: "the term 'chemical substance' means any organic or inorganic substance of a particular molecular identity, including – (i) any combination of such substances occurring in whole or in part as a result of a chemical reaction or occurring in nature"

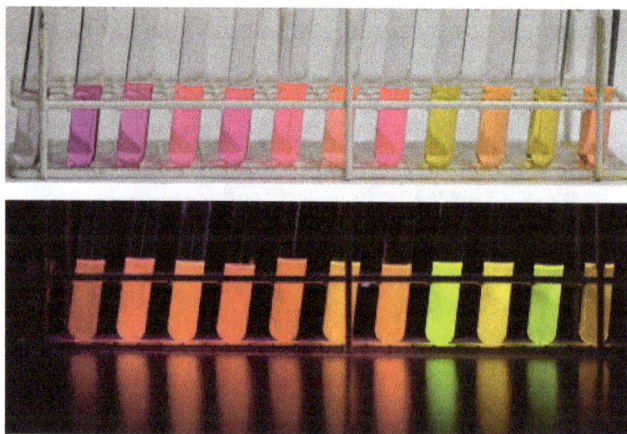

Colors of a single chemical (Nile red) in different solvents, under visible and UV light, showing how the chemical interacts dynamically with its solvent environment.

In geology, substances of uniform composition are called minerals, while physical mixtures (aggregates) of several minerals (different substances) are defined as rocks. Many minerals, however, mutually dissolve into solid solutions, such that a single rock is a uniform substance despite being a mixture in stoichiometric terms. Feldspars are a common example: anorthoclase is an alkali aluminium silicate, where the alkali metal is interchangeably either sodium or potassium.

In law, "chemical substances" may include both pure substances and mixtures with a defined composition or manufacturing process. For example, the EU regulation REACH defines "monoconstituent substances", "multiconstituent substances" and "substances of unknown or variable composition". The latter two consist of multiple chemical substances; however, their identity can be established either by direct chemical analysis or reference to a single manufacturing process. For example, charcoal is an extremely complex, partially polymeric mixture that can be defined by its manufacturing process. Therefore, although the exact chemical identity is unknown, identification can be made to a sufficient accuracy. The CAS index also includes mixtures.

Polymers almost always appear as mixtures of molecules of multiple molar masses, each of which could be considered a separate chemical substance. However, the polymer may be defined by a known precursor or reaction(s) and the molar mass distribution. For example, polyethylene is a mixture of very long chains of $-CH_2-$ repeating units, and is generally sold in several molar mass distributions, LDPE, MDPE, HDPE and UHMWPE.

History

The concept of a "chemical substance" became firmly established in the late eighteenth century after work by the chemist Joseph Proust on the composition of some pure chemical compounds such as basic copper carbonate. He deduced that, "All samples of a compound have the same composition; that is, all samples have the same proportions, by mass, of the elements present in the compound." This is now known as the law of constant composition. Later with the advancement of methods for chemical synthesis particularly in the realm of organic chemistry; the discovery of many more chemical elements and new techniques in the realm of analytical chemistry used for isolation and purification of elements and compounds from chemicals that led to the establishment of modern chemistry, the concept was defined as is found in most chemistry textbooks. However, there are some controversies regarding this definition mainly because the large number of chemical substances reported in chemistry literature need to be indexed.

Isomerism caused much consternation to early researchers, since isomers have exactly the same composition, but differ in configuration (arrangement) of the atoms. For example, there was much speculation for the chemical identity of benzene, until the correct structure was described by Friedrich August Kekulé. Likewise, the idea of stereoisomerism - that atoms have rigid three-dimensional structure and can thus form isomers that differ only in their three-dimensional arrangement - was another crucial step in understanding the concept of distinct chemical substances. For example, tartaric acid has three distinct isomers, a pair of diastereomers with one diastereomer forming two enantiomers.

Chemical Elements

An element is a chemical substance that is made up of a particular kind of atoms and hence cannot be broken down or transformed by a chemical reaction into a different element, though it can be transmutated into another element through a nuclear reaction. This is so, because all of the atoms in a sample of an element have the same number of protons, though they may be different isotopes, with differing numbers of neutrons.

Native sulfur crystals. Sulfur occurs naturally as elemental sulfur, in sulfide and sulfate minerals and in hydrogen sulfide.

As of 2012, there are 118 known elements, about 80 of which are stable – that is, they do not change by radioactive decay into other elements. Some elements can occur as more than a single chemical substance (allotropes). For instance, oxygen exists as both diatomic oxygen (O_2) and ozone (O_3). The majority of elements are classified as metals. These are elements with a characteristic lustre such as iron, copper, and gold. Metals typically conduct electricity and heat well, and they are malleable and ductile. Around a dozen elements, such as carbon, nitrogen, and oxygen, are classified as non-metals. Non-metals lack the metallic properties described above, they also have a high electronegativity and a tendency to form negative ions. Certain elements such as silicon sometimes resemble metals and sometimes resemble non-metals, and are known as metalloids.

Chemical Compounds

A pure chemical compound is a chemical substance that is composed of a particular set of molecules or ions. Two or more elements combined into one substance through a chemical reaction form a chemical compound. All compounds are substances, but not all substances are compounds.

Potassium ferricyanide is a compound of potassium, iron, carbon and nitrogen; although it contains cyanide anions, it does not release them and is nontoxic.

A chemical compound can be either atoms bonded together in molecules or crystals in which atoms, molecules or ions form a crystalline lattice. Compounds based primarily on carbon and hydrogen atoms are called organic compounds, and all others are called inorganic compounds. Compounds containing bonds between carbon and a metal are called organometallic compounds.

Compounds in which components share electrons are known as covalent compounds. Compounds consisting of oppositely charged ions are known as ionic compounds, or salts.

In organic chemistry, there can be more than one chemical compound with the same composition and molecular weight. Generally, these are called isomers. Isomers usually have substantially different chemical properties, may be isolated and do not spontaneously convert to each other. A common example is glucose vs. fructose. The former is an aldehyde, the latter is a ketone. Their interconversion requires either enzymatic or acid-base catalysis. However, there are also tautomers, where isomerization occurs spontaneously, such that a pure substance cannot be isolated into its tautomers. A common example is glucose, which has open-chain and ring forms. One cannot manufacture pure open-chain glucose because glucose spontaneously cyclizes to the hemiacetal form. Materials may also comprise other entities such as polymers. These may be inorganic or organic and sometimes a combination of inorganic and organic.

Substances Versus Mixtures

All matter consists of various elements and chemical compounds, but these are often intimately mixed together. Mixtures contain more than one chemical substance, and they do not have a fixed composition. In principle, they can be separated into the component substances by purely mechanical processes. Butter, soil and wood are common examples of mixtures.

Cranberry glass, while it looks homogeneous, is a *mixture* consisting of glass and gold colloidal particles of ca. 40 nm diameter, which give it a red color.

Grey iron metal and yellow sulfur are both chemical elements, and they can be mixed together in any ratio to form a yellow-grey mixture. No chemical process occurs, and the material can be identified as a mixture by the fact that the sulfur and the iron can be separated by a mechanical process, such as using a magnet to attract the iron away from the sulfur.

In contrast, if iron and sulfur are heated together in a certain ratio (1 atom of iron for each atom of sulfur, or by weight, 56 grams (1 mol) of iron to 32 grams (1 mol) of sulfur), a chemical reaction takes place and a new substance is formed, the compound iron(II) sulfide, with chemical formula FeS. The resulting compound has all the properties of a chemical substance and is not a mixture. Iron(II) sulfide has its own distinct properties such as melting point and solubility, and the two elements cannot be separated using normal mechanical processes; a magnet will be unable to recover the iron, since there is no metallic iron present in the compound.

Chemicals Versus Chemical Substances

While the term *chemical substance* is a precise technical term that is synonymous with "chemical" for professional chemists, the meaning of the word *chemical* varies for non-chemists within the English speaking world or those using English. For industries, government and society in general in some countries, the word *chemical* includes a wider class of substances that contain many mixtures of such chemical substances, often finding application in many vocations. In countries that require a list of ingredients in products, the "chemicals" listed would be equated with "chemical substances".

Within the chemical industry, manufactured "chemicals" are chemical substances, which can be classified by production volume into bulk chemicals, fine chemicals and chemicals found in research only:

- Bulk chemicals are produced in very large quantities, usually with highly optimized continuous processes and to a relatively low price.

- Fine chemicals are produced at a high cost in small quantities for special low-volume applications such as biocides, pharmaceuticals and speciality chemicals for technical applications.

- Research chemicals are produced individually for research, such as when searching for synthetic routes or screening substances for pharmaceutical activity. In effect, their price per gram is very high, although they are not sold.

The cause of the difference in production volume is the complexity of the molecular structure of the chemical. Bulk chemicals are usually much less complex. While fine chemicals may be more complex, many of them are simple enough to be sold as "building blocks" in the synthesis of more complex molecules targeted for single use, as named above. The *production* of a chemical includes not only its synthesis but also its purification to eliminate by-products and impurities involved in the synthesis. The last step in production should be the analysis of batch lots of chemicals in order to identify and quantify the percentages of impurities for the buyer of the chemicals. The required purity and analysis depends on the application, but higher tolerance of impurities is usually expected in the production of bulk chemicals. Thus, the user of the chemical in the US might choose between the bulk or "technical grade" with higher amounts of impurities or a much purer "pharmaceutical grade" (labeled "USP", United States Pharmacopeia).

Naming and Indexing

Every chemical substance has one or more systematic names, usually named according to the IUPAC rules for naming. An alternative system is used by the Chemical Abstracts Service (CAS).

Many compounds are also known by their more common, simpler names, many of which predate the systematic name. For example, the long-known sugar glucose is now systematically named 6-(hydroxymethyl)oxane-2,3,4,5-tetrol. Natural products and pharmaceuticals are also given simpler names, for example the mild pain-killer Naproxen is the more common name for the chemical compound (S)-6-methoxy-α-methyl-2-naphthaleneacetic acid.

Chemists frequently refer to chemical compounds using chemical formulae or molecular structure of the compound. There has been a phenomenal growth in the number of chemical compounds being synthesized (or isolated), and then reported in the scientific literature by professional chemists around the world. An enormous number of chemical compounds are possible through the chemical combination of the known chemical elements. As of May 2011, about sixty million chemical compounds are known. The names of many of these compounds are often nontrivial and hence not very easy to remember or cite accurately. Also it is difficult to keep the track of them in the literature. Several international organizations like IUPAC and CAS have initiated steps to make such tasks easier. CAS provides the abstracting services of the chemical literature, and provides a numerical identifier, known as CAS registry number to each chemical substance that has been reported in the chemical literature (such as chemistry journals and patents). This information is compiled as a database and is popularly known as the Chemical substances index. Other computer-friendly systems that have been developed for substance information, are: SMILES and the International Chemical Identifier or InChI.

Identification of a typical chemical substance					
Common name	Systematic name	Chemical formula	Chemical structure	CAS registry number	InChI
alcohol, or ethyl alcohol	ethanol	C_2H_5OH	$\diagup\!\diagdown$ OH	[64-17-5]	1/C2H6O/c1-2-3/ h3H,2H2,1H3

Isolation, Purification, Characterization, and Identification

Often a pure substance needs to be isolated from a mixture, for example from a natural source (where a sample often contains numerous chemical substances) or after a chemical reaction (which often give mixtures of chemical substances).

Chemical Element

A chemical element or element is a species of atoms having the same number of protons in their atomic nuclei (i.e. the same atomic number, Z). There are 118 elements that have been identified, of which the first 94 occur naturally on Earth with the remaining 24 being synthetic elements. There are 80 elements that have at least one stable isotope and 38 that have exclusively radioactive isotopes, which decay over time into other elements. Iron is the most abundant element (by mass) making up Earth, while oxygen is the most common element in the crust of Earth.

Chemical elements constitute all of the ordinary matter of the universe. However astronomical observations suggest that ordinary observable matter is only approximately 15% of the matter in the universe: the remainder is dark matter, the composition of which is unknown, but it is not composed of chemical elements. The two lightest elements, hydrogen and helium were mostly formed in the Big Bang and are the most common elements in the universe. The next three elements (lithium, beryllium and boron) were formed mostly by cosmic ray spallation, and are thus more rare than those that follow. Formation of elements with from six to twenty six protons occurred and continues to occur in main sequence stars via stellar nucleosynthesis. The high abundance of oxygen, silicon, and iron on Earth reflects their common production in such stars. Elements with greater than twenty-six protons are formed by supernova nucleosynthesis in supernovae, which, when they explode, blast these elements far into space as supernova remnants, where they may become incorporated into planets when they are formed.

The term "element" is used for a kind of atoms with a given number of protons (regardless of whether they are or they are not ionized or chemically bonded, e.g. hydrogen in water) as well as for a pure chemical substance consisting of a single element (e.g. hydrogen gas). For the second meaning, the terms "elementary substance" and "simple substance" have been suggested, but they have not gained much acceptance in the English-language chemical literature, whereas in some other languages their equivalent is widely used (e.g. French *corps simple*, Russian *простое вещество*). One element can form multiple substances different by their structure; they are called allotropes of the element.

When different elements are chemically combined, with the atoms held together by chemical bonds, they form chemical compounds. Only a minority of elements are found uncombined as relatively pure minerals. Among the more common of such "native elements" are copper, silver, gold, carbon (as coal, graphite, or diamonds), and sulfur. All but a few of the most inert elements, such as noble gases and noble metals, are usually found on Earth in chemically combined form, as chemical compounds. While about 32 of the chemical elements occur on Earth in native uncombined forms, most of these occur as mixtures. For example, atmospheric air is primarily a mixture of nitrogen, oxygen, and argon, and native solid elements occur in alloys, such as that of iron and nickel.

The history of the discovery and use of the elements began with primitive human societies that found native elements like carbon, sulfur, copper and gold. Later civilizations extracted elemental copper, tin, lead and iron from their ores by smelting, using charcoal. Alchemists and chemists subsequently identified many more, with almost all of the naturally-occurring elements becoming known by 1900.

The properties of the chemical elements are summarized on the periodic table, which organizes the elements by increasing atomic number into rows ("periods") in which the columns ("groups") share recurring ("periodic") physical and chemical properties. Save for unstable radioactive elements with short half-lives, all of the elements are available industrially, most of them in high degrees of purity.

Description

The lightest chemical elements are hydrogen and helium, both created by Big Bang nucleosynthesis during the first 20 minutes of the universe in a ratio of around 3:1 by mass (or 12:1 by number of atoms), along with tiny traces of the next two elements, lithium and beryllium. Almost all other elements found in nature were made by various natural methods of nucleosynthesis. On Earth, small amounts of new atoms are naturally produced in nucleogenic reactions, or in cosmogenic processes, such as cosmic ray spallation. New atoms are also naturally produced on Earth as radiogenic daughter isotopes of ongoing radioactive decay processes such as alpha decay, beta decay, spontaneous fission, cluster decay, and other rarer modes of decay.

Of the 94 naturally occurring elements, those with atomic numbers 1 through 82 each have at least one stable isotope (except for technetium, element 43 and promethium, element 61, which have no stable isotopes). Isotopes considered stable are those for which no radioactive decay has yet been observed. Elements with atomic numbers 83 through 94 are unstable to the point that radioactive decay of all isotopes can be detected. Some of these elements, notably bismuth (atomic number 83), thorium (atomic number 90), and uranium (atomic number 92), have one or more isotopes with half-lives long enough to survive as remnants of the explosive stellar nucleosynthesis that produced the heavy elements before the formation of our solar system. At over 1.9×10^{19} years, over a billion times longer than the current estimated age of the universe, bismuth-209 (atomic number 83) has the longest known alpha decay half-life of any naturally occurring element, and is almost always considered on par with the 80 stable elements. The very heaviest elements (those beyond plutonium, element 94) undergo radioactive decay with half-lives so short that they are not found in nature and must be synthesized.

As of 2010, there are 118 known elements (in this context, "known" means observed well enough, even from just a few decay products, to have been differentiated from other elements). Of these 118 elements, 94 occur naturally on Earth. Six of these occur in extreme trace quantities: technetium, atomic number 43; promethium, number 61; astatine, number 85; francium, number 87; neptunium, number 93; and plutonium, number 94. These 94 elements have been detected in the universe at large, in the spectra of stars and also supernovae, where short-lived radioactive elements are newly being made. The first 94 elements have been detected directly on Earth as primordial nuclides present from the formation of the solar system, or as naturally-occurring fission or transmutation products of uranium and thorium.

The remaining 24 heavier elements, not found today either on Earth or in astronomical spectra, have been produced artificially: these are all radioactive, with very short half-lives; if any atoms of these elements were present at the formation of Earth, they are extremely likely, to the point of certainty, to have already decayed, and if present in novae, have been in quantities too small to have been noted. Technetium was the first purportedly non-naturally occurring element synthesized, in 1937, although trace amounts of technetium have since been found in nature (and also the element may have been discovered naturally in 1925). This pattern of artificial production and later natural discovery has been repeated with several other radioactive naturally-occurring rare elements.

Lists of the elements are available by name, by symbol, by atomic number, by density, by melting point, and by boiling point as well as ionization energies of the elements.

The nuclides of stable and radioactive elements are also available as a list of nuclides, sorted by length of half-life for those that are unstable. One of the most convenient, and certainly the most traditional presentation of the elements, is in the form of the periodic table, which groups together elements with similar chemical properties (and usually also similar electronic structures).

Atomic Number

The atomic number of an element is equal to the number of protons in each atom, and defines the element. For example, all carbon atoms contain 6 protons in their atomic nucleus; so the atomic number of carbon is 6. Carbon atoms may have different numbers of neutrons; atoms of the same element having different numbers of neutrons are known as isotopes of the element.

The number of protons in the atomic nucleus also determines its electric charge, which in turn determines the number of electrons of the atom in its non-ionized state. The electrons are placed into atomic orbitals that determine the atom's various chemical properties. The number of neutrons in a nucleus usually has very little effect on an element's chemical properties (except in the case of hydrogen and deuterium). Thus, all carbon isotopes have nearly identical chemical properties because they all have six protons and six electrons, even though carbon atoms may, for example, have 6 or 8 neutrons. That is why the atomic number, rather than mass number or atomic weight, is considered the identifying characteristic of a chemical element.

The symbol for atomic number is Z.

Isotopes

Isotopes are atoms of the same element (that is, with the same number of protons in their atomic nucleus), but having *different* numbers of neutrons. Most (66 of 94) naturally occurring elements have more than one stable isotope. Thus, for example, there are three main isotopes of carbon. All carbon atoms have 6 protons in the nucleus, but they can have either 6, 7, or 8 neutrons. Since the mass numbers of these are 12, 13 and 14 respectively, the three isotopes of carbon are known as carbon-12, carbon-13, and carbon-14, often abbreviated to ^{12}C, ^{13}C, and ^{14}C. Carbon in everyday life and in chemistry is a mixture of ^{12}C (about 98.9%), ^{13}C (about 1.1%) and about 1 atom per trillion of ^{14}C.

Except in the case of the isotopes of hydrogen (which differ greatly from each other in relative mass—enough to cause chemical effects), the isotopes of a given element are chemically nearly indistinguishable.

All of the elements have some isotopes that are radioactive (radioisotopes), although not all of these radioisotopes occur naturally. The radioisotopes typically decay into other elements upon radiating an alpha or beta particle. If an element has isotopes that are not radioactive, these are termed "stable" isotopes. All of the known stable isotopes occur naturally. The many radioisotopes that are not found in nature have been characterized after being artificially made. Certain elements have no stable isotopes and are composed *only* of radioactive isotopes: specifically the elements without any stable isotopes are technetium (atomic number 43), promethium (atomic number 61), and all observed elements with atomic numbers greater than 82.

Of the 80 elements with at least one stable isotope, 26 have only one single stable isotope. The mean number of stable isotopes for the 80 stable elements is 3.1 stable isotopes per element. The largest number of stable isotopes that occur for a single element is 10 (for tin, element 50).

Isotopic Mass and Atomic Mass

The mass number of an element, A, is the number of nucleons (protons and neutrons) in the atomic nucleus. Different isotopes of a given element are distinguished by their mass numbers, which are conventionally written as a superscript on the left hand side of the atomic symbol (e.g., ^{238}U). The mass number is always a simple whole number and has units of "nucleons." An example of a referral to a mass number is "magnesium-24," which is an atom with 24 nucleons (12 protons and 12 neutrons).

Whereas the mass number simply counts the total number of neutrons and protons and is thus a natural (or whole) number, the atomic mass of a single atom is a real number for the mass of a particular isotope of the element, the unit being u. In general, when expressed in u it differs in value slightly from the mass number for a given nuclide (or isotope) since the mass of the protons and neutrons is not exactly 1 u, since the electrons contribute a lesser share to the atomic mass as neutron number exceeds proton number, and (finally) because of the nuclear binding energy. For example, the atomic mass of chlorine-35 to five significant digits is 34.969 u and that of chlorine-37 is 36.966 u. However, the atomic mass in u of each isotope is quite close to its simple mass number (always within 1%). The only isotope whose atomic mass is exactly a natural number is ^{12}C, which by definition has a mass of exactly 12, because u is defined as 1/12 of the mass of a free neutral carbon-12 atom in the ground state.

The relative atomic mass (historically and commonly also called "atomic weight") of an element is the *average* of the atomic masses of all the chemical element's isotopes as found in a particular environment, weighted by isotopic abundance, relative to the atomic mass unit (u). This number may be a fraction that is *not* close to a whole number, due to the averaging process. For example, the relative atomic mass of chlorine is 35.453 u, which differs greatly from a whole number due to being made of an average of 76% chlorine-35 and 24% chlorine-37. Whenever a relative atomic mass value differs by more than 1% from a whole number, it is due to this averaging effect resulting from significant amounts of more than one isotope being naturally present in the sample of the element in question.

Chemically Pure and Isotopically Pure

Chemists and nuclear scientists have different definitions of a *pure element*. In chemistry, a pure element means a substance whose atoms all (or in practice almost all) have the same atomic number, or number of protons. Nuclear scientists, however, define a pure element as one that consists of only one stable isotope.

For example, a copper wire is 99.99% chemically pure if 99.99% of its atoms are copper, with 29 protons each. However it is not isotopically pure since ordinary copper consists of two stable isotopes, 69% ^{63}Cu and 31% ^{65}Cu, with different numbers of neutrons. However, a pure gold ingot would be both chemically and isotopically pure, since ordinary gold consists only of one isotope, ^{197}Au.

Allotropes

Atoms of chemically pure elements may bond to each other chemically in more than one way, allowing the pure element to exist in multiple structures (spatial arrangements of atoms), known as allotropes, which differ in their properties. For example, carbon can be found as diamond, which has a tetrahedral structure around each carbon atom; graphite, which has layers of carbon atoms with a hexagonal structure stacked on top of each other; graphene, which is a single layer of graphite that is very strong; fullerenes, which have nearly spherical shapes; and carbon nanotubes, which are tubes with a hexagonal structure (even these may differ from each other in electrical properties). The ability of an element to exist in one of many structural forms is known as 'allotropy'.

The standard state, also known as reference state, of an element is defined as its thermodynamically most stable state at 1 bar at a given temperature (typically at 298.15 K). In thermochemistry, an element is defined to have an enthalpy of formation of zero in its standard state. For example, the reference state for carbon is graphite, because the structure of graphite is more stable than that of the other allotropes.

Properties

Several kinds of descriptive categorizations can be applied broadly to the elements, including consideration of their general physical and chemical properties, their states of matter under familiar conditions, their melting and boiling points, their densities, their crystal structures as solids, and their origins.

General Properties

Several terms are commonly used to characterize the general physical and chemical properties of the chemical elements. A first distinction is between metals, which readily conduct electricity, nonmetals, which do not, and a small group, (the *metalloids*), having intermediate properties and often behaving as semiconductors.

A more refined classification is often shown in colored presentations of the periodic table. This system restricts the terms "metal" and "nonmetal" to only certain of the more broadly defined metals and nonmetals, adding additional terms for certain sets of the more broadly viewed metals and nonmetals. The version of this classification used in the periodic tables presented here includes: actinides, alkali metals, alkaline earth metals, halogens, lanthanides, transition metals, post-transition metals, metalloids, polyatomic nonmetals, diatomic nonmetals, and noble gases. In this system, the alkali metals, alkaline earth metals, and transition metals, as well as the lanthanides and the actinides, are special groups of the metals viewed in a broader sense. Similarly, the polyatomic nonmetals, diatomic nonmetals and the noble gases are nonmetals viewed in the broader sense. In some presentations, the halogens are not distinguished, with astatine identified as a metalloid and the others identified as nonmetals.

States of Matter

Another commonly used basic distinction among the elements is their state of matter (phase), whether solid, liquid, or gas, at a selected standard temperature and pressure (STP). Most of the elements are solids at conventional temperatures and atmospheric pressure, while several are gases. Only bromine and mercury are liquids at 0 degrees Celsius (32 degrees Fahrenheit) and normal atmospheric pressure; caesium and gallium are solids at that temperature, but melt at 28.4 °C (83.2 °F) and 29.8 °C (85.6 °F), respectively.

Melting and Boiling Points

Melting and boiling points, typically expressed in degrees Celsius at a pressure of one atmosphere, are commonly used in characterizing the various elements. While known for most elements, either or both of these measurements is still undetermined for some of the radioactive elements available in only tiny quantities. Since helium remains a liquid even at absolute zero at atmospheric pressure, it has only a boiling point, and not a melting point, in conventional presentations.

Densities

The density at a selected standard temperature and pressure (STP) is frequently used in characterizing the elements. Density is often expressed in grams per cubic centimeter (g/cm^3). Since several elements are gases at commonly encountered temperatures, their densities are usually stated for their gaseous forms; when liquefied or solidified, the gaseous elements have densities similar to those of the other elements.

When an element has allotropes with different densities, one representative allotrope is typically selected in summary presentations, while densities for each allotrope can be stated where more detail is provided. For example, the three familiar allotropes of carbon (amorphous carbon, graphite, and diamond) have densities of 1.8–2.1, 2.267, and 3.515 g/cm^3, respectively.

Crystal Structures

The elements studied to date as solid samples have eight kinds of crystal structures: cubic, body-centered cubic, face-centered cubic, hexagonal, monoclinic, orthorhombic, rhombohedral, and tetragonal. For some of the synthetically produced transuranic elements, available samples have been too small to determine crystal structures.

Occurrence and Origin on Earth

Chemical elements may also be categorized by their origin on Earth, with the first 94 considered naturally occurring, while those with atomic numbers beyond 94 have only been produced artificially as the synthetic products of man-made nuclear reactions.

Of the 94 naturally occurring elements, 84 are considered primordial and either stable or weakly radioactive. The remaining 10 naturally occurring elements possess half lives too short for them to have been present at the beginning of the Solar System, and are therefore considered transient elements. (Plutonium is usually also considered a transient element because primordial plutonium has by now decayed to almost undetectable traces.) Of these 10 transient elements, 5 (polonium, radon, radium, actinium, and protactinium) are relatively common decay products of thorium, uranium, and plutonium. The remaining 6 transient elements (technetium, promethium, astatine, francium, neptunium, and plutonium) occur only rarely, as products of rare decay modes or nuclear reaction processes involving uranium or other heavy elements.

Elements with atomic numbers 1 through 40 are all stable, while those with atomic numbers 41 through 82 (except technetium and promethium) are metastable. The half-lives of these metastable "theoretical radionuclides" are so long (at least 100 million times longer than the estimated age of the universe) that their radioactive decay has yet to be detected by experiment. Elements with atomic numbers 83 through 94 are unstable to the point that their radioactive decay can be detected. Three of these elements, bismuth (element 83), thorium (element 90), and uranium (element 92) have one or more isotopes with half-lives long enough to survive as remnants of the explosive stellar nucleosynthesis that produced the heavy elements before the formation of our solar system. For example, at over 1.9×10^{19} years, over a billion times longer than the current estimated age of the universe, bismuth-209 has the longest known alpha decay half-life of any naturally occurring element. The very heaviest 24 elements (those beyond plutonium, element 94) undergo radioactive decay with short half-lives and cannot be produced as daughters of longer-lived elements, and thus they do not occur in nature at all.

The Periodic Table

The properties of the chemical elements are often summarized using the periodic table, which powerfully and elegantly organizes the elements by increasing atomic number into rows ("periods") in which the columns ("groups") share recurring ("periodic") physical and chemical properties. The current standard table contains 118 confirmed elements as of 10 April 2010.

Although earlier precursors to this presentation exist, its invention is generally credited to the Russian chemist Dmitri Mendeleev in 1869, who intended the table to illustrate recurring trends in the properties of the elements. The layout of the table has been refined and extended over time as new elements have been discovered and new theoretical models have been developed to explain chemical behavior.

Use of the periodic table is now ubiquitous within the academic discipline of chemistry, providing an extremely useful framework to classify, systematize and compare all the many different forms of chemical behavior. The table has also found wide application in physics, geology, biology, materials science, engineering, agriculture, medicine, nutrition, environmental health, and astronomy. Its principles are especially important in chemical engineering.

Nomenclature and Symbols

The various chemical elements are formally identified by their unique atomic numbers, by their accepted names, and by their symbols.

Atomic Numbers

The known elements have atomic numbers from 1 through 118, conventionally presented as Arabic numerals. Since the elements can be uniquely sequenced by atomic number, conventionally from lowest to highest (as in a periodic table), sets of elements are sometimes specified by such notation as "through", "beyond", or "from ... through", as in "through iron", "beyond uranium", or "from lanthanum through lutetium". The terms "light" and "heavy" are sometimes also used informally to indicate relative atomic numbers (not densities), as in "lighter than carbon" or "heavier than lead", although technically the weight or mass of atoms of an element (their atomic weights or atomic masses) do not always increase monotonically with their atomic numbers.

Element Names

The naming of various substances now known as elements precedes the atomic theory of matter, as names were given locally by various cultures to various minerals, metals, compounds, alloys, mixtures, and other materials, although at the time it was not known which chemicals were elements and which compounds. As they were identified as elements, the existing names for anciently-known elements (e.g., gold, mercury, iron) were kept in most countries. National differences emerged over the names of elements either for convenience, linguistic niceties, or nationalism. For a few illustrative examples: German speakers use "Wasserstoff" (water substance) for "hydrogen", "Sauerstoff" (acid substance) for "oxygen" and "Stickstoff" (smothering substance) for "nitrogen", while English and some romance languages use "sodium" for "natrium" and "potassium" for "kalium", and the French, Italians, Greeks, Portuguese and Poles prefer "azote/azot/azoto" (from roots meaning "no life") for "nitrogen".

For purposes of international communication and trade, the official names of the chemical elements both ancient and more recently recognized are decided by the International Union of Pure and Applied Chemistry (IUPAC), which has decided on a sort of international English language, drawing on traditional English names even when an element's chemical symbol is based on a Latin or other traditional word, for example adopting "gold" rather than "aurum" as the name for the 79th element (Au). IUPAC prefers the British spellings "aluminium" and "caesium" over the U.S. spellings "aluminum" and "cesium", and the U.S. "sulfur" over the British "sulphur". However, elements that are practical to sell in bulk in many countries often still have locally used national names, and countries whose national language does not use the Latin alphabet are likely to use the IUPAC element names.

According to IUPAC, chemical elements are not proper nouns in English; consequently, the full name of an element is not routinely capitalized in English, even if derived from a proper noun, as in californium and einsteinium. Isotope names of chemical elements are also uncapitalized if written out, *e.g.*, carbon-12 or uranium-235. Chemical element *symbols* (such as Cf for californium and Es for einsteinium), are always capitalized.

In the second half of the twentieth century, physics laboratories became able to produce nuclei of chemical elements with half-lives too short for an appreciable amount of them to exist at any time. These are also named by IUPAC, which generally adopts the name chosen by the discoverer. This practice can lead to the controversial question of which research group actually discovered an element, a question that delayed the naming of elements with atomic number of 104 and higher for a considerable amount of time.

Precursors of such controversies involved the nationalistic namings of elements in the late 19th century. For example, *lutetium* was named in reference to Paris, France. The Germans were reluctant to relinquish naming rights to the French, often calling it *cassiopeium*. Similarly, the British discoverer of *niobium* originally named it *columbium*, in reference to the New World. It was used extensively as such by American publications prior to the international standardization (in 1950).

Chemical Symbols

Specific Chemical Elements

Before chemistry became a science, alchemists had designed arcane symbols for both metals and common compounds. These were however used as abbreviations in diagrams or procedures; there was no concept of atoms combining to form molecules.

With his advances in the atomic theory of matter, John Dalton devised his own simpler symbols, based on circles, to depict molecules.

The current system of chemical notation was invented by Berzelius. In this typographical system, chemical symbols are not mere abbreviations—though each consists of letters of the Latin alphabet. They are intended as universal symbols for people of all languages and alphabets.

The first of these symbols were intended to be fully universal. Since Latin was the common language of science at that time, they were abbreviations based on the Latin names of metals. Cu comes from Cuprum, Fe comes from Ferrum, Ag from Argentum. The symbols were not followed by a period (full stop) as with abbreviations. Later chemical elements were also assigned unique chemical symbols, based on the name of the element, but not necessarily in English. For example, sodium has the chemical symbol 'Na' after the Latin *natrium*. The same applies to "W" (wolfram) for tungsten, "Fe" (ferrum) for iron, "Hg" (hydrargyrum) for mercury, "Sn" (stannum) for tin, "K" (kalium) for potassium, "Au" (aurum) for gold, "Ag" (argentum) for silver, "Pb" (plumbum) for lead, "Cu" (cuprum) for copper, and "Sb" (stibium) for antimony.

Chemical symbols are understood internationally when element names might require translation. There have sometimes been differences in the past. For example, Germans in the past have used "J" (for the alternate name Jod) for iodine, but now use "I" and "Iod".

The first letter of a chemical symbol is always capitalized, as in the preceding examples, and the subsequent letters, if any, are always lower case (small letters). Thus, the symbols for californium or einsteinium are Cf and Es.

General Chemical Symbols

There are also symbols in chemical equations for groups of chemical elements, for example in comparative formulas. These are often a single capital letter, and the letters are reserved and not used for names of specific elements. For example, an "X" indicates a variable group (usually a halogen) in a class of compounds, while "R" is a radical, meaning a compound structure such as a hydrocarbon chain. The letter "Q" is reserved for "heat" in a chemical reaction. "Y" is also often used as a general chemical symbol, although it is also the symbol of yttrium. "Z" is also frequently used as a general variable group. "E" is used in organic chemistry to denote an electron-withdrawing group or an electrophile; similarly "Nu" denotes a nucleophile. "L" is used to represent a general ligand in inorganic and organometallic chemistry. "M" is also often used in place of a general metal.

At least two additional, two-letter generic chemical symbols are also in informal usage, "Ln" for any lanthanide element and "An" for any actinide element. "Rg" was formerly used for any rare gas element, but the group of rare gases has now been renamed noble gases and the symbol "Rg" has now been assigned to the element roentgenium.

Isotope Symbols

Isotopes are distinguished by the atomic mass number (total protons and neutrons) for a particular isotope of an element, with this number combined with the pertinent element's symbol. IUPAC prefers that isotope symbols be written in superscript notation when practical, for example ^{12}C and ^{235}U. However, other notations, such as carbon-12 and uranium-235, or C-12 and U-235, are also used.

As a special case, the three naturally occurring isotopes of the element hydrogen are often specified as H for ^{1}H (protium), D for ^{2}H (deuterium), and T for ^{3}H (tritium). This convention is easier to use in chemical equations, replacing the need to write out the mass number for each atom. For example, the formula for heavy water may be written D_2O instead of ^2H_2O.

Origin of the Elements

Only about 4% of the total mass of the universe is made of atoms or ions, and thus represented by chemical elements. This fraction is about 15% of the total matter, with the remainder of the matter (85%) being dark matter. The nature of dark matter is unknown, but it is not composed of atoms of chemical elements because it contains no protons, neutrons, or electrons. (The remaining non-matter part of the mass of the universe is composed of the even more mysterious dark energy).

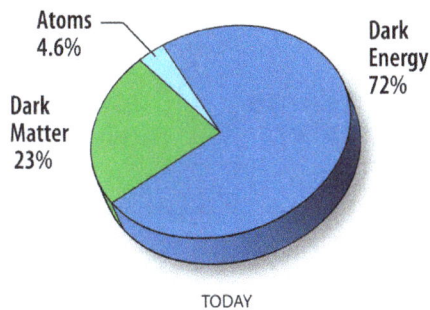

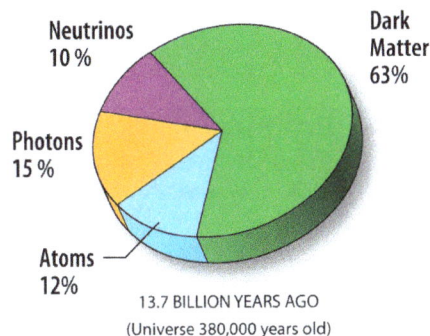

Estimated distribution of dark matter and dark energy in the universe. Only the fraction of the mass and energy in the universe labeled "atoms" is composed of chemical elements.

The universe's 94 naturally occurring chemical elements are thought to have been produced by at least four cosmic processes. Most of the hydrogen and helium in the universe was produced primordially in the first few minutes of the Big Bang. Three recurrently occurring later processes are thought to have produced the remaining elements. Stellar nucleosynthesis, an ongoing process, produces all elements from carbon through iron in atomic number, but little lithium, beryllium, or boron. Elements heavier in atomic number than iron, as heavy as uranium and plutonium, are produced by explosive nucleosynthesis in supernovas and other cataclysmic cosmic events. Cosmic ray spallation (fragmentation) of carbon, nitrogen, and oxygen is important to the production of lithium, beryllium and boron.

During the early phases of the Big Bang, nucleosynthesis of hydrogen nuclei resulted in the production of hydrogen-1 (protium, ^{1}H) and helium-4 (^{4}He), as well as a smaller amount of deuterium (^{2}H) and very minuscule amounts (on the order of 10^{-10}) of lithium and beryllium. Even smaller amounts of boron may have been produced in the Big Bang, since it has been observed in some very old stars, while carbon has not. It is generally agreed that no heavier elements than boron were produced in the Big Bang. As a result, the primordial abundance of atoms (or ions) consisted of roughly 75% ^{1}H, 25% ^{4}He, and 0.01% deuterium, with only tiny traces of lithium, beryllium, and perhaps boron. Subsequent enrichment of galactic halos occurred due to stellar nucleosynthesis and supernova nucleosynthesis. However, the element abundance in intergalactic space can still closely resemble primordial conditions, unless it has been enriched by some means.

Periodic table showing the cosmogenic origin of each element in the Big Bang, or in large or small stars. Small stars can produce certain elements up to sulfur, by the alpha process. Supernovae are needed to produce "heavy" elements (those beyond iron and nickel) rapidly by neutron buildup, in the r-process. Certain large stars slowly produce other elements heavier than iron, in the s-process; these may then be blown into space in the off-gassing of planetary nebulae

On Earth (and elsewhere), trace amounts of various elements continue to be produced from other elements as products of natural transmutation processes. These include some produced by cosmic rays or other nuclear reactions, and others produced as decay products of long-lived primordial nuclides. For example, trace (but detectable) amounts of carbon-14 (^{14}C) are continually produced in the atmosphere by cosmic rays impacting nitrogen atoms, and argon-40 (^{40}Ar) is continually produced by the decay of primordially occurring but unstable potassium-40 (^{40}K). Also, three primordially occurring but radioactive actinides, thorium, uranium, and plutonium, decay through a series of recurrently produced but unstable radioactive elements such as radium and radon, which are transiently present in any sample of these metals or their ores or compounds. Three other radioactive elements, technetium, promethium, and neptunium, occur only incidentally in natural materials, produced as individual atoms by natural fission of the nuclei of various heavy elements or in other rare nuclear processes.

Human technology has produced various additional elements beyond these first 94, with those through atomic number 118 now known.

Abundance

The following graph (note log scale) shows the abundance of elements in our solar system. The table shows the twelve most common elements in our galaxy (estimated spectroscopically), as measured in parts per million, by mass. Nearby galaxies that have evolved along similar lines have a corresponding enrichment of elements heavier than hydrogen and helium. The more distant galaxies are being viewed as they appeared in the past, so their abundances of elements appear closer to the primordial mixture. As physical laws and processes appear common throughout the visible universe, however, scientist expect that these galaxies evolved elements in similar abundance.

The abundance of elements in the Solar System is in keeping with their origin from nucleosynthesis in the Big Bang and a number of progenitor supernova stars. Very abundant hydrogen and helium are products of the Big Bang, but the next three elements are rare since they had little time to form in the Big Bang and are not made in stars (they are, however, produced in small quantities by the breakup of heavier elements in interstellar dust, as a result of impact by cosmic rays). Beginning with carbon, elements are produced in stars by buildup from alpha particles (helium nuclei), resulting in an alternatingly larger abundance of elements with even atomic numbers (these are also more stable). In general, such elements up to iron are made in large stars in the process of becoming supernovas. Iron-56 is particularly common, since it is the most stable element that can easily be made from alpha particles (being a product of decay of radioactive nickel-56, ultimately made from 14 helium nuclei). Elements heavier than iron are made in energy-absorbing processes in large stars, and their abundance in the universe (and on Earth) generally decreases with their atomic number.

The abundance of the chemical elements on Earth varies from air to crust to ocean, and in various types of life. The abundance of elements in Earth's crust differs from that in the Solar system (as seen in the Sun and heavy planets like Jupiter) mainly in selective loss of the very lightest elements (hydrogen and helium) and also volatile neon, carbon (as hydrocarbons), nitrogen and sulfur, as a result of solar heating in the early formation of the solar system. Oxygen, the most abundant Earth element by mass, is retained on Earth by combination with silicon. Aluminum at 8% by mass is more common in the Earth's crust than in the universe and solar system, but the composition of the far more bulky mantle, which has magnesium and iron in place of aluminum (which occurs there only at 2% of mass) more closely mirrors the elemental composition of the solar system, save for the noted loss of volatile elements to space, and loss of iron which has migrated to the Earth's core.

The composition of the human body, by contrast, more closely follows the composition of seawater—save that the human body has additional stores of carbon and nitrogen necessary to form the proteins and nucleic acids, together with phosphorus in the nucleic acids and energy transfer molecule adenosine triphosphate (ATP) that occurs in the cells of all living organisms. Certain kinds of organisms require particular additional elements, for example the magnesium in chlorophyll in green plants, the calcium in mollusc shells, or the iron in the hemoglobin in vertebrate animals' red blood cells.

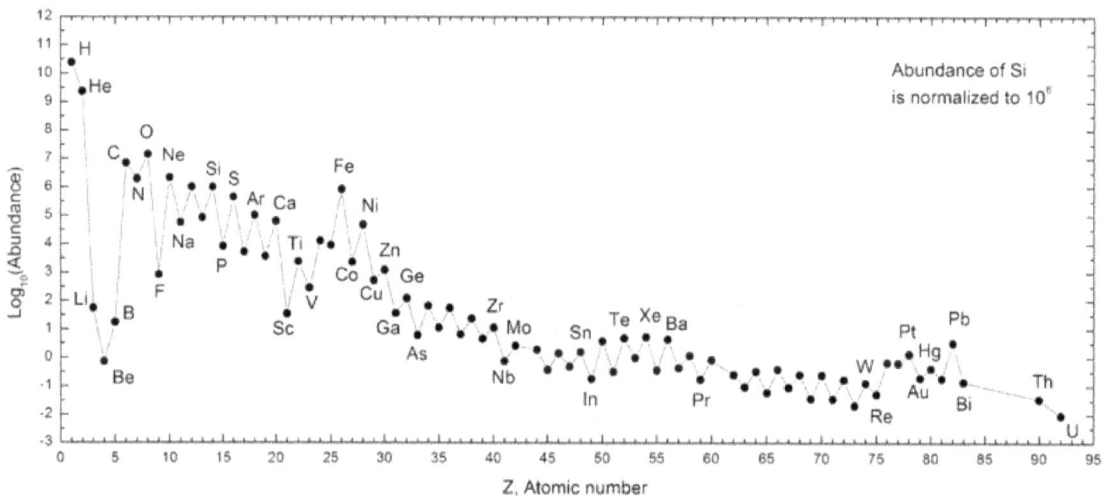

Abundances of the chemical elements in the Solar system. Hydrogen and helium are most common, from the Big Bang. The next three elements (Li, Be, B) are rare because they are poorly synthesized in the Big Bang and also in stars. The two general trends in the remaining stellar-produced elements are: (1) an alternation of abundance in elements as they have even or odd atomic numbers (the Oddo-Harkins rule), and (2) a general decrease in abundance as elements become heavier. Iron is especially common because it represents the minimum energy nuclide that can be made by fusion of helium in supernovae.

Elements in our galaxy	Parts per million by mass
Hydrogen	739,000
Helium	240,000
Oxygen	10,400
Carbon	4,600
Neon	1,340
Iron	1,090
Nitrogen	960
Silicon	650
Magnesium	580
Sulfur	440
Potassium	210
Nickel	100

Nutritional elements in the periodic table

H																	He
Li	Be											B	C	N	O	F	Ne
Na	Mg											Al	Si	P	S	Cl	Ar
K	Ca	Sc	Ti	V	Cr	Mn	Fe	Co	Ni	Cu	Zn	Ga	Ge	As	Se	Br	Kr
Rb	Sr	Y	Zr	Nb	Mo	Tc	Ru	Rh	Pd	Ag	Cd	In	Sn	Sb	Te	I	Xe
Cs	Ba	La	*	Hf	Ta	W	Re	Os	Ir	Pt	Au	Hg	Tl	Pb	Bi	Po	At	Rn
Fr	Ra	Ac	**	Rf	Db	Sg	Bh	Hs	Mt	Ds	Rg	Cn	Uut	Fl	Ms	Lv	Ts	Og

	*	Ce	Pr	Nd	Pm	Sm	Eu	Gd	Tb	Dy	Ho	Er	Tm	Yb	Lu
	**	Th	Pa	U	Np	Pu	Am	Cm	Bk	Cf	Es	Fm	Md	No	Lr

The four organic basic elements

Quantity elements

Essential trace elements

Suggested function from deprivation effects or active metabolic handling, but no clearly-identified biochemical function in humans

Limited circumstantial evidence for trace benefits or biological action in mammals

No evidence for biological action in mammals, but essential in some lower organisms

History

ОПЫТЪ СИСТЕМЫ ЭЛЕМЕНТОВЪ,

ОСНОВАННОЙ НА ИХЪ АТОМНОМЪ ВѢСѢ И ХИМИЧЕСКОМЪ СХОДСТВѢ.

```
                              Ti=50     Zr=90    ?=180.
                              V=51      Nb=94    Ta=182.
                              Cr=52     Mo=96    W=186.
                              Mn=55     Rh=104,4 Pt=197,1.
                              Fe=56     Ru=104,4 Ir=198.
                           Ni=Co=59     Pd=106,6 Os=199.
            H=1               Cu=63,4    Ag=108   Hg=200.
                  Be= 9,4 Mg=24  Zn=65,2 Cd=112
                  B=11    Al=27,3   ?=68  Ur=116   Au=197?
                  C=12    Si=28     ?=70  Sn=118
                  N=14    P=31    As=75   Sb=122   Bi=210?
                  O=16    S=32    Se=79,4 Te=128?
                  F=19    Cl=35,5 Br=80   I=127
            Li=7  Na=23   K=39    Rb=85,4 Cs=133   Tl=204.
                          Ca=40   Sr=87,6 Ba=137   Pb=207.
                           ?=45   Ce=92
                          ?Er=56  La=94
                          ?Yt=60  Di=95
                          ?In=75,6 Th=118?
```

Д. Менделѣевъ

Mendeleev's 1869 periodic table: *An experiment on a system of elements. Based on their atomic weights and chemical similarities.*

Evolving Definitions

The concept of an "element" as an undivisible substance has developed through three major historical phases: Classical definitions (such as those of the ancient Greeks), chemical definitions, and atomic definitions.

Classical Definitions

Ancient philosophy posited a set of classical elements to explain observed patterns in nature. These *elements* originally referred to *earth, water, air* and *fire* rather than the chemical elements of modern science.

The term 'elements' (*stoicheia*) was first used by the Greek philosopher Plato in about 360 BCE in his dialogue Timaeus, which includes a discussion of the composition of inorganic and organic bodies and is a speculative treatise on chemistry. Plato believed the elements introduced a century earlier by Empedocles were composed of small polyhedral forms: tetrahedron (fire), octahedron (air), icosahedron (water), and cube (earth).

Aristotle, c. 350 BCE, also used the term *stoicheia* and added a fifth element called aether, which formed the heavens. Aristotle defined an element as:

Element – one of those bodies into which other bodies can decompose, and that itself is not capable of being divided into other.

Chemical Definitions

In 1661, Robert Boyle proposed his theory of corpuscularism which favoured the analysis of matter as constituted by irreducible units of matter (atoms) and, choosing to side with neither Aristotle's view of the four elements nor Paracelsus' view of three fundamental elements, left open the question of the number of elements. The first modern list of chemical elements was given in Antoine Lavoisier's 1789 *Elements of Chemistry*, which contained thirty-three elements, including light and caloric. By 1818, Jöns Jakob Berzelius had determined atomic weights for forty-five of the forty-nine then-accepted elements. Dmitri Mendeleev had sixty-six elements in his periodic table of 1869.

Dmitri Mendeleev

From Boyle until the early 20th century, an element was defined as a pure substance that could not be decomposed into any simpler substance. Put another way, a chemical element cannot be transformed into other chemical elements by chemical processes. Elements during this time were generally distinguished by their atomic weights, a property measurable with fair accuracy by available analytical techniques.

Atomic Definitions

The 1913 discovery by English physicist Henry Moseley that the nuclear charge is the physical basis for an atom's atomic number, further refined when the nature of protons and neutrons became appreciated, eventually led to the current definition of an element based on atomic number (number of protons per atomic nucleus). The use of atomic numbers, rather than atomic weights, to distinguish elements has greater predictive value (since these numbers are integers), and also resolves some ambiguities in the chemistry-based view due to varying properties of isotopes and allotropes within the same element. Currently, IUPAC defines an element to exist if it has isotopes with a lifetime longer than the 10^{-14} seconds it takes the nucleus to form an electronic cloud.

Henry Moseley

By 1914, seventy-two elements were known, all naturally occurring. The remaining naturally occurring elements were discovered or isolated in subsequent decades, and various additional elements have also been produced synthetically, with much of that work pioneered by Glenn T. Seaborg. In 1955, element 101 was discovered and named mendelevium in honor of D.I. Mendeleev, the first to arrange the elements in a periodic manner. Most recently, the synthesis of element 118 was reported in October 2006, and the synthesis of element 117 was reported in April 2010.

Discovery and Recognition of Various Elements

Ten materials familiar to various prehistoric cultures are now known to be chemical elements: Carbon, copper, gold, iron, lead, mercury, silver, sulfur, tin, and zinc. Three additional materials now accepted as elements, arsenic, antimony, and bismuth, were recognized as distinct substances prior to 1500 AD. Phosphorus, cobalt, and platinum were isolated before 1750.

Most of the remaining naturally occurring chemical elements were identified and characterized by 1900, including:

- Such now-familiar industrial materials as aluminium, silicon, nickel, chromium, magnesium, and tungsten

- Reactive metals such as lithium, sodium, potassium, and calcium

- The halogens fluorine, chlorine, bromine, and iodine

- Gases such as hydrogen, oxygen, nitrogen, helium, argon, and neon

- Most of the rare-earth elements, including cerium, lanthanum, gadolinium, and neodymium.

- The more common radioactive elements, including uranium, thorium, radium, and radon

Elements isolated or produced since 1900 include:

- The three remaining undiscovered regularly occurring stable natural elements: hafnium, lutetium, and rhenium

- Plutonium, which was first produced synthetically in 1940 by Glenn T. Seaborg, but is now also known from a few long-persisting natural occurrences

- The three incidentally occurring natural elements (neptunium, promethium, and technetium), which were all first produced synthetically but later discovered in trace amounts in certain geological samples

- Three scarce decay products of uranium or thorium, (astatine, francium, and protactinium), and

- Various synthetic transuranic elements, beginning with americium and curium

Recently Discovered Elements

The first transuranium element (element with atomic number greater than 92) discovered was neptunium in 1940. Since 1999 claims for the discovery of new elements have been considered by the IUPAC/IUPAP Joint Working Party. As of January 2016, all 118 elements have been confirmed as discovered by IUPAC. The discovery of element 112 was acknowledged in 2009, and the name *copernicium* and the atomic symbol *Cn* were suggested for it. The name and symbol were officially endorsed by IUPAC on 19 February 2010. The heaviest element that is believed to have been synthesized to date is element 118, ununoctium, on 9 October 2006, by the Flerov Laboratory of Nuclear Reactions in Dubna, Russia. Element 117 was the latest element claimed to be discovered, in 2009. IUPAC officially recognized flerovium and livermorium, elements 114 and 116, in June 2011 and approved their names in May 2012. In December 2015, IUPAC recognized elements 113, 115, 117 and 118, and announced the elements' proposed final names on 8 June 2016. The names, nihonium (113, Nh), moscovium (115, Mc), tennessine (117, Ts), and oganesson (118, Og), are expected to be approved by the end of 2016.

List of the 118 Known Chemical Elements

The following sortable table shows the 118 known chemical elements.

- Atomic number, name, and symbol all serve independently as unique identifiers.

- Names are those accepted by IUPAC; provisional names for recently produced elements not yet formally named are in parentheses.

- Group, period, and block refer to an element's position in the periodic table. Group numbers here show the currently accepted numbering; for older alternate numberings, see Group (periodic table).

- State of matter *(solid, liquid,* or *gas)* applies at standard temperature and pressure conditions (STP).

- Occurrence distinguishes naturally occurring elements, categorized as either *primordial* or *transient* (from decay), and additional *synthetic* elements that have been produced technologically, but are not known to occur naturally.

- Description summarizes an element's properties using the broad categories commonly presented in periodic tables: Actinide, alkali metal, alkaline earth metal, lanthanide, post-transition metal, metalloid, noble gas, polyatomic or diatomic nonmetal, and transition metal.

List of elements								
Atomic no.	Name	Symbol	Group	Period	Block	State at STP	Occurrence	Description
1	Hydrogen	H	1	1	s	Gas	Primordial	Diatomic nonmetal
2	Helium	He	18	1	s	Gas	Primordial	Noble gas
3	Lithium	Li	1	2	s	Solid	Primordial	Alkali metal
4	Beryllium	Be	2	2	s	Solid	Primordial	Alkaline earth metal
5	Boron	B	13	2	p	Solid	Primordial	Metalloid
6	Carbon	C	14	2	p	Solid	Primordial	Polyatomic nonmetal
7	Nitrogen	N	15	2	p	Gas	Primordial	Diatomic nonmetal
8	Oxygen	O	16	2	p	Gas	Primordial	Diatomic nonmetal
9	Fluorine	F	17	2	p	Gas	Primordial	Diatomic nonmetal
10	Neon	Ne	18	2	p	Gas	Primordial	Noble gas
11	Sodium	Na	1	3	s	Solid	Primordial	Alkali metal
12	Magnesium	Mg	2	3	s	Solid	Primordial	Alkaline earth metal

13	Aluminium	Al	13	3	p	Solid	Primordial	Post-transition metal
14	Silicon	Si	14	3	p	Solid	Primordial	Metalloid
15	Phosphorus	P	15	3	p	Solid	Primordial	Polyatomic nonmetal
16	Sulfur	S	16	3	p	Solid	Primordial	Polyatomic nonmetal
17	Chlorine	Cl	17	3	p	Gas	Primordial	Diatomic nonmetal
18	Argon	Ar	18	3	p	Gas	Primordial	Noble gas
19	Potassium	K	1	4	s	Solid	Primordial	Alkali metal
20	Calcium	Ca	2	4	s	Solid	Primordial	Alkaline earth metal
21	Scandium	Sc	3	4	d	Solid	Primordial	Transition metal
22	Titanium	Ti	4	4	d	Solid	Primordial	Transition metal
23	Vanadium	V	5	4	d	Solid	Primordial	Transition metal
24	Chromium	Cr	6	4	d	Solid	Primordial	Transition metal
25	Manganese	Mn	7	4	d	Solid	Primordial	Transition metal
26	Iron	Fe	8	4	d	Solid	Primordial	Transition metal
27	Cobalt	Co	9	4	d	Solid	Primordial	Transition metal
28	Nickel	Ni	10	4	d	Solid	Primordial	Transition metal
29	Copper	Cu	11	4	d	Solid	Primordial	Transition metal
30	Zinc	Zn	12	4	d	Solid	Primordial	Transition metal
31	Gallium	Ga	13	4	p	Solid	Primordial	Post-transition metal
32	Germanium	Ge	14	4	p	Solid	Primordial	Metalloid
33	Arsenic	As	15	4	p	Solid	Primordial	Metalloid
34	Selenium	Se	16	4	p	Solid	Primordial	Polyatomic nonmetal
35	Bromine	Br	17	4	p	Liquid	Primordial	Diatomic nonmetal
36	Krypton	Kr	18	4	p	Gas	Primordial	Noble gas
37	Rubidium	Rb	1	5	s	Solid	Primordial	Alkali metal
38	Strontium	Sr	2	5	s	Solid	Primordial	Alkaline earth metal
39	Yttrium	Y	3	5	d	Solid	Primordial	Transition metal
40	Zirconium	Zr	4	5	d	Solid	Primordial	Transition metal
41	Niobium	Nb	5	5	d	Solid	Primordial	Transition metal
42	Molybdenum	Mo	6	5	d	Solid	Primordial	Transition metal
43	Technetium	Tc	7	5	d	Solid	Transient	Transition metal
44	Ruthenium	Ru	8	5	d	Solid	Primordial	Transition metal
45	Rhodium	Rh	9	5	d	Solid	Primordial	Transition metal
46	Palladium	Pd	10	5	d	Solid	Primordial	Transition metal
47	Silver	Ag	11	5	d	Solid	Primordial	Transition metal

48	Cadmium	Cd	12	5	d	Solid	Primordial	Transition metal
49	Indium	In	13	5	p	Solid	Primordial	Post-transition metal
50	Tin	Sn	14	5	p	Solid	Primordial	Post-transition metal
51	Antimony	Sb	15	5	p	Solid	Primordial	Metalloid
52	Tellurium	Te	16	5	p	Solid	Primordial	Metalloid
53	Iodine	I	17	5	p	Solid	Primordial	Diatomic nonmetal
54	Xenon	Xe	18	5	p	Gas	Primordial	Noble gas
55	Caesium	Cs	1	6	s	Solid	Primordial	Alkali metal
56	Barium	Ba	2	6	s	Solid	Primordial	Alkaline earth metal
57	Lanthanum	La	3	6	f	Solid	Primordial	Lanthanide
58	Cerium	Ce	3	6	f	Solid	Primordial	Lanthanide
59	Praseodymium	Pr	3	6	f	Solid	Primordial	Lanthanide
60	Neodymium	Nd	3	6	f	Solid	Primordial	Lanthanide
61	Promethium	Pm	3	6	f	Solid	Transient	Lanthanide
62	Samarium	Sm	3	6	f	Solid	Primordial	Lanthanide
63	Europium	Eu	3	6	f	Solid	Primordial	Lanthanide
64	Gadolinium	Gd	3	6	f	Solid	Primordial	Lanthanide
65	Terbium	Tb	3	6	f	Solid	Primordial	Lanthanide
66	Dysprosium	Dy	3	6	f	Solid	Primordial	Lanthanide
67	Holmium	Ho	3	6	f	Solid	Primordial	Lanthanide
68	Erbium	Er	3	6	f	Solid	Primordial	Lanthanide
69	Thulium	Tm	3	6	f	Solid	Primordial	Lanthanide
70	Ytterbium	Yb	3	6	f	Solid	Primordial	Lanthanide
71	Lutetium	Lu	3	6	d	Solid	Primordial	Lanthanide
72	Hafnium	Hf	4	6	d	Solid	Primordial	Transition metal
73	Tantalum	Ta	5	6	d	Solid	Primordial	Transition metal
74	Tungsten	W	6	6	d	Solid	Primordial	Transition metal
75	Rhenium	Re	7	6	d	Solid	Primordial	Transition metal
76	Osmium	Os	8	6	d	Solid	Primordial	Transition metal
77	Iridium	Ir	9	6	d	Solid	Primordial	Transition metal
78	Platinum	Pt	10	6	d	Solid	Primordial	Transition metal
79	Gold	Au	11	6	d	Solid	Primordial	Transition metal
80	Mercury	Hg	12	6	d	Liquid	Primordial	Transition metal
81	Thallium	Tl	13	6	p	Solid	Primordial	Post-transition metal
82	Lead	Pb	14	6	p	Solid	Primordial	Post-transition metal

83	Bismuth	Bi	15	6	p	Solid	Primordial	Post-transition metal
84	Polonium	Po	16	6	p	Solid	Transient	Post-transition metal
85	Astatine	At	17	6	p	Solid	Transient	Metalloid
86	Radon	Rn	18	6	p	Gas	Transient	Noble gas
87	Francium	Fr	1	7	s	Solid	Transient	Alkali metal
88	Radium	Ra	2	7	s	Solid	Transient	Alkaline earth metal
89	Actinium	Ac	3	7	f	Solid	Transient	Actinide
90	Thorium	Th	3	7	f	Solid	Primordial	Actinide
91	Protactinium	Pa	3	7	f	Solid	Transient	Actinide
92	Uranium	U	3	7	f	Solid	Primordial	Actinide
93	Neptunium	Np	3	7	f	Solid	Transient	Actinide
94	Plutonium	Pu	3	7	f	Solid	Transient	Actinide
95	Americium	Am	3	7	f	Solid	Synthetic	Actinide
96	Curium	Cm	3	7	f	Solid	Synthetic	Actinide
97	Berkelium	Bk	3	7	f	Solid	Synthetic	Actinide
98	Californium	Cf	3	7	f	Solid	Synthetic	Actinide
99	Einsteinium	Es	3	7	f	Solid	Synthetic	Actinide
100	Fermium	Fm	3	7	f		Synthetic	Actinide
101	Mendelevium	Md	3	7	f		Synthetic	Actinide
102	Nobelium	No	3	7	f		Synthetic	Actinide
103	Lawrencium	Lr	3	7	d		Synthetic	Actinide
104	Rutherfordium	Rf	4	7	d		Synthetic	Transition metal
105	Dubnium	Db	5	7	d		Synthetic	Transition metal
106	Seaborgium	Sg	6	7	d		Synthetic	Transition metal
107	Bohrium	Bh	7	7	d		Synthetic	Transition metal
108	Hassium	Hs	8	7	d		Synthetic	Transition metal
109	Meitnerium	Mt	9	7	d		Synthetic	
110	Darmstadtium	Ds	10	7	d		Synthetic	
111	Roentgenium	Rg	11	7	d		Synthetic	
112	Copernicium	Cn	12	7	d		Synthetic	Transition metal
113	(Ununtrium)	(Uut)	13	7	p		Synthetic	
114	Flerovium	Fl	14	7	p		Synthetic	Post-transition metal
115	(Ununpentium)	(Uup)	15	7	p		Synthetic	
116	Livermorium	Lv	16	7	p		Synthetic	
117	(Ununseptium)	(Uus)	17	7	p		Synthetic	
118	(Ununoctium)	(Uuo)	18	7	p		Synthetic	

References

- Wolf, Jakob (15 January 2010). Schnellkurs HGB-Jahresabschluss: Das neue Bilanzrecht: Richtig vorgehen — erfolgreich umstellen. Walhalla Fachverlag. p. 90. ISBN 978-3-8029-3436-0.

- Oerter, Robert (2006). The Theory of Almost Everything: The Standard Model, the Unsung Triumph of Modern Physics. Penguin. p. 223. ISBN 978-0-452-28786-0.

- "IUPAC Is Naming The Four New Elements Nihonium, Moscovium, Tennessine, And Oganesson". IUPAC. 2016-06-08. Retrieved 2016-06-08.

- Dumé, B (23 April 2003). "Bismuth breaks half-life record for alpha decay". Physicsworld.com. Bristol, England: Institute of Physics. Retrieved 14 July 2015.

- Harvard–Smithsonian Center for Astrophysics. "ORIGIN OF HEAVY ELEMENTS". cfa.harvard.edu. Retrieved 26 February 2013.

- Joachim Schummer. "Coping with the Growth of Chemical Knowledge: Challenges for Chemistry Documentation, Education, and Working Chemists". Rz.uni-karlsruhe.de. Retrieved 2013-06-06.

- Los Alamos National Laboratory (2011). "Periodic Table of Elements: Oxygen". Los Alamos, New Mexico: Los Alamos National Security, LLC. Retrieved 7 May 2011.

2

Branches of Chemical Engineering

Chemical engineering is an interdisciplinary subject. It has numerous branches; some of these are biochemical engineering, electrochemical engineering and chemical process modeling. Biochemical engineering mainly deals with designing and construction of unit processes whereas electrochemical engineering deals with the functions of electrochemical phenomena. This text will provide a glimpse of the related fields of chemical engineering briefly.

Biochemical Engineering

Biochemical engineering is a branch of chemical engineering that mainly deals with the design and construction of unit processes that involve biological organisms or molecules, such as bioreactors. Its applications are in the petrochemical industry, food, pharmaceutical, biotechnology, and water treatment industries.

Bioreactor

Electrochemical Engineering

Electrochemical engineering is the branch of engineering dealing with the technological applications of electrochemical phenomena, such as synthesis of chemicals, electrowinning and refining of metals, batteries and fuel cells, sensors, surface modification by electrodeposition and etching, separations, and corrosion. It is an overlap between electrochemistry and chemical engineering. One of the pioneers of this field of engineering was Charles Frederick Burgess.

More than 6% of the electrical energy is consumed by electrochemical operations in the USA.

History

This branch of engineering emerged gradually from chemical engineering. The works of Wagner (1962) and Levich (1962) influenced the emergence of electrochemical engineering, because their work inspired so many others. Several individuals, including Tobias, Ibl, and Hine, established engineering training centers and, with their colleagues, developed important experimental and theoretical methods of study.

Scope

Electrochemical engineering combines the study of heterogeneous charge transfer at electrode/electrolyte interphases with the development of practical materials and processes. Fundamental considerations include electrode materials and the kinetics of redox species. The development of the technology involves the study of the electrochemical reactors, their potential and current distribution, mass transport conditions, hydrodynamics, geometry and components as well as the quantification of its overall performance in terms of reaction yield, conversion efficiency, and energy efficiency. Industrial developments require further reactor and process design, fabrication methods, testing and product development.

Applications

Electrochemical engineering finds applications in chemical synthesis, ion/organics removal, deposition of fils of metals and semiconductors, sensors and monitoring and energy storage and conversion.

3

Processes in Chemical Engineering

Chemical reactions are processes that transform one set of chemical substances into another. Chemical processes are processes that change chemicals or chemical compounds. Other themes included in this section include unit operation, process design, chemical process modeling and process integration. This chapter discusses the methods of chemical engineering in a critical manner providing key analysis to the subject matter.

Chemical Process

In a scientific sense, a chemical process is a method or means of somehow changing one or more chemicals or chemical compounds. Such a chemical process can occur by itself or be caused by an outside force, and involves a chemical reaction of some sort. In an "engineering" sense, a chemical process is a method intended to be used in manufacturing or on an industrial scale to change the composition of chemical(s) or material(s), usually using technology similar or related to that used in chemical plants or the chemical industry.

Neither of these definitions is exact in the sense that one can always tell definitively what is a chemical process and what is not; they are practical definitions. There is also significant overlap in these two definition variations. Because of the inexactness of the definition, chemists and other scientists use the term "chemical process" only in a general sense or in the engineering sense. However, in the "process (engineering)" sense, the term "chemical process" is used extensively. The rest of the article will cover the engineering type of chemical process.

Although this type of chemical process may sometimes involve only one step, often multiple steps, referred to as unit operations, are involved. In a plant, each of the unit operations commonly occur in individual vessels or sections of the plant called units. Often, one or more chemical reactions are involved, but other ways of changing chemical (or material) composition may be used, such as mixing or separation processes. The process steps may be sequential in time or sequential in space along a stream of flowing or moving material. For a given amount of a feed (input) material or product (output) material, an expected amount of material can be determined at key steps in the process from empirical data and material balance calculations. These amounts can be scaled up or down to suit the desired capacity or operation of a particular chemical plant built for such a process. More than one chemical plant may use the same chemical process, each plant perhaps at differently scaled capacities. Chemical processes like distillation and crystallization go back to alchemy in Alexandria, Egypt.

Such chemical processes can be illustrated generally as block flow diagrams or in more detail as process flow diagrams. Block flow diagrams show the units as blocks and the streams flowing between them as connecting lines with arrowheads to show direction of flow.

In addition to chemical plants for producing chemicals, chemical processes with similar technology and equipment are also used in oil refining and other refineries, natural gas processing, polymer and pharmaceutical manufacturing, food processing, and water and wastewater treatment.

Unit Processing in Chemical Process

Unit processing is the basic processing in chemical engineering. Together with unit operations it forms the main principle of the varied chemical industries. Each genre of unit processing follows the same chemical law much as each genre of unit operations follows the same physical law.

Chemical engineering unit processing consists of the following important processes:

- Oxidation
- Reduction
- Hydrogenation
- Dehydrogenation
- Hydrolysis
- Hydration
- Dehydration
- Halogenation
- Nitrification
- Sulfonation
- Ammoniation
- Alkaline fusion
- Alkylation
- Dealkylation
- Esterification
- Polymerization

- Polycondensation

- Catalysis

Academic Research Institutes in Process Chemistry

Institute of Process Research & Development, University of Leeds

Redox

Redox (short for reduction–oxidation reaction) is a chemical reaction in which the oxidation states of atoms are changed. Any such reaction involves both a reduction process and a complementary oxidation process, two key concepts involved with electron transfer processes. Redox reactions include all chemical reactions in which atoms have their oxidation state changed; in general, redox reactions involve the transfer of electrons between chemical species. The chemical species from which the electron is stripped is said to have been oxidized, while the chemical species to which the electron is added is said to have been reduced. It can be explained in simple terms:

- Oxidation is the *loss* of electrons or an *increase* in oxidation state by a molecule, atom, or ion.

- Reduction is the *gain* of electrons or a *decrease* in oxidation state by a molecule, atom, or ion.

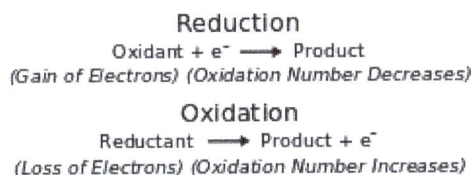

Reduction
Oxidant + e⁻ ⟶ Product
(Gain of Electrons) (Oxidation Number Decreases)

Oxidation
Reductant ⟶ Product + e⁻
(Loss of Electrons) (Oxidation Number Increases)

The two parts of a redox reaction

Rusting iron

A bonfire

As an example, during the combustion of wood, oxygen from the air is reduced, transferring electrons from the carbon. Although oxidation reactions are commonly associated with the formation of oxides from oxygen molecules, oxygen is not necessarily included in such reactions, as other chemical species can serve the same function.

The reaction can occur relatively slowly, as in the case of rust, or more quickly, as in the case of fire. There are simple redox processes, such as the oxidation of carbon to yield carbon dioxide (CO_2) or the reduction of carbon by hydrogen to yield methane (CH_4), and more complex processes such as the oxidation of glucose ($C_6H_{12}O_6$) in the human body.

Etymology

"Redox" is a combination of "reduction" and "oxidation".

The word *oxidation* originally implied reaction with oxygen to form an oxide, since dioxygen (O_2 (g)) was historically the first recognized oxidizing agent. Later, the term was expanded to encompass oxygen-like substances that accomplished parallel chemical reactions. Ultimately, the meaning was generalized to include all processes involving loss of electrons.

The word *reduction* originally referred to the loss in weight upon heating a metallic ore such as a metal oxide to extract the metal. In other words, ore was "reduced" to metal. Antoine Lavoisier (1743–1794) showed that this loss of weight was due to the loss of oxygen as a gas. Later, scientists realized that the metal atom gains electrons in this process. The meaning of *reduction* then became generalized to include all processes involving gain of electrons. Even though "reduction" seems counter-intuitive when speaking of the gain of electrons, it might help to think of reduction as the loss of oxygen, which was its historical meaning. Since electrons are negatively charged, it is also helpful to think of this as reduction in electrical charge.

The electrochemist John Bockris has used the words *electronation* and *deelectronation* to describe reduction and oxidation processes respectively when they occur at electrodes. These words are analogous to protonation and deprotonation, but they have not been widely adopted by chemists.

The term "hydrogenation" could be used instead of reduction, since hydrogen is the reducing agent in a large number of reactions, especially in organic chemistry and biochemistry. But, unlike oxidation, which has been generalized beyond its root element, hydrogenation has maintained its specific connection to reactions that *add* hydrogen to another substance (e.g., the hydrogenation of unsaturated fats into saturated fats, $R-CH=CH-R + H_2 \rightarrow R-CH_2-CH_2-R$). The word "redox" was first used in 1928.

Definitions

The processes of oxidation and reduction occur simultaneously and cannot happen independently of one another, similar to the acid–base reaction. The oxidation alone and the reduction alone are each called a *half-reaction*, because two half-reactions always occur together to form a whole reaction. When writing half-reactions, the gained or lost electrons are typically included explicitly in order that the half-reaction be balanced with respect to electric charge.

Though sufficient for many purposes, these general descriptions are not precisely correct. Although oxidation and reduction properly refer to *a change in oxidation state* — the actual transfer of electrons may never occur. The oxidation state of an atom is the fictitious charge that an atom would have if all bonds between atoms of different elements were 100% ionic. Thus, oxidation is best defined as an *increase in oxidation state*, and reduction as a *decrease in oxidation state*. In practice, the transfer of electrons will always cause a change in oxidation state, but there are many reactions that are classed as "redox" even though no electron transfer occurs (such as those involving covalent bonds).

Oxidizing and Reducing Agents

In redox processes, the reductant transfers electrons to the oxidant. Thus, in the reaction, the reductant or *reducing agent* loses electrons and is oxidized, and the oxidant or *oxidizing agent* gains electrons and is reduced. The pair of an oxidizing and reducing agent that are involved in a particular reaction is called a redox pair. A redox couple is a reducing species and its corresponding oxidizing form, e.g., Fe^{2+}/Fe^{3+}.

Oxidizers

Substances that have the ability to oxidize other substances (cause them to lose electrons) are said to be oxidative or oxidizing and are known as oxidizing agents, oxidants, or oxidizers. That is, the oxidant (oxidizing agent) removes electrons from another substance, and is thus itself reduced. And, because it "accepts" electrons, the oxidizing agent is also called an electron acceptor. Oxygen is the quintessential oxidizer.

The international pictogram for oxidising chemicals.

Oxidants are usually chemical substances with elements in high oxidation states (e.g., H2O2, MnO−4, CrO3, Cr2O2−7, OsO4), or else highly electronegative elements (O_2, F_2, Cl_2, Br_2) that can gain extra electrons by oxidizing another substance.

Reducers

Substances that have the ability to reduce other substances (cause them to gain electrons) are said to be reductive or reducing and are known as reducing agents, reductants, or reducers. The reductant (reducing agent) transfers electrons to another substance, and is thus itself oxidized. And, because it "donates" electrons, the reducing agent is also called an electron donor. Electron donors can also form charge transfer complexes with electron acceptors.

Reductants in chemistry are very diverse. Electropositive elemental metals, such as lithium, sodium, magnesium, iron, zinc, and aluminium, are good reducing agents. These metals donate or *give away* electrons readily. *Hydride transfer reagents*, such as $NaBH_4$ and $LiAlH_4$, are widely used in organic chemistry, primarily in the reduction of carbonyl compounds to alcohols. Another method of reduction involves the use of hydrogen gas (H_2) with a palladium, platinum, or nickel catalyst. These *catalytic reductions* are used primarily in the reduction of carbon-carbon double or triple bonds.

Standard Electrode Potentials (Reduction Potentials)

Each half-reaction has a *standard electrode potential* (Eocell), which is equal to the potential difference or voltage at equilibrium under standard conditions of an electrochemical cell in which the cathode reaction is the half-reaction considered, and the anode is a standard hydrogen electrode where hydrogen is oxidized:

$$\tfrac{1}{2}\, H_2 \rightarrow H^+ + e^-.$$

The electrode potential of each half-reaction is also known as its *reduction potential Eo red*, or potential when the half-reaction takes place at a cathode. The reduction potential is a measure of the tendency of the oxidizing agent to be reduced. Its value is zero for $H^+ + e^- \rightarrow \frac{1}{2} H_2$ by definition, positive for oxidizing agents stronger than H^+ (e.g., +2.866 V for F_2) and negative for oxidizing agents that are weaker than H^+ (e.g., −0.763 V for Zn^{2+}).

For a redox reaction that takes place in a cell, the potential difference is:

$$Eo_{cell} = Eo_{cathode} - Eo_{anode}$$

However, the potential of the reaction at the anode was sometimes expressed as an *oxidation potential*:

$$Eo_{ox} = -Eo_{red}.$$

The oxidation potential is a measure of the tendency of the reducing agent to be oxidized, but does not represent the physical potential at an electrode. With this notation, the cell voltage equation is written with a plus sign

$$Eo_{cell} = Eo_{red}(\text{cathode}) + Eo_{ox}(\text{anode})$$

Examples of Redox Reactions

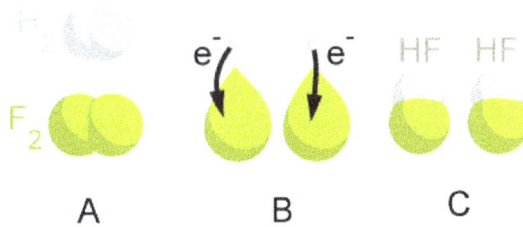

Illustration of a redox reaction

A good example is the reaction between hydrogen and fluorine in which hydrogen is being oxidized and fluorine is being reduced:

$$H_2 + F2 \rightarrow 2 \, HF$$

We can write this overall reaction as two half-reactions:

the oxidation reaction:

$$H_2 \rightarrow 2 \, H^+ + 2 \, e^-$$

and the reduction reaction:

$$F_2 + 2 \, e^- \rightarrow 2 \, F^-$$

Analyzing each half-reaction in isolation can often make the overall chemical process clearer. Because there is no net change in charge during a redox reaction, the number of electrons in excess in the oxidation reaction must equal the number consumed by the reduction reaction.

Elements, even in molecular form, always have an oxidation state of zero. In the first half-reaction, hydrogen is oxidized from an oxidation state of zero to an oxidation state of +1. In the second half-reaction, fluorine is reduced from an oxidation state of zero to an oxidation state of −1.

When adding the reactions together the electrons are canceled:

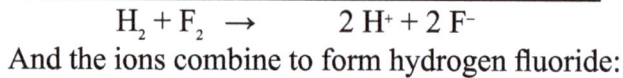

$$H_2 \rightarrow \quad 2\,H^+ + 2\,e^-$$
$$F_2 + 2\,e^- \rightarrow \quad 2\,F^-$$

$$\overline{}$$
$$H_2 + F_2 \rightarrow \quad 2\,H^+ + 2\,F^-$$

And the ions combine to form hydrogen fluoride:

$$2\,H^+ + 2\,F^- \rightarrow 2\,HF$$

The overall reaction is:

$$H_2 + F_2 \rightarrow 2\,HF$$

Metal Displacement

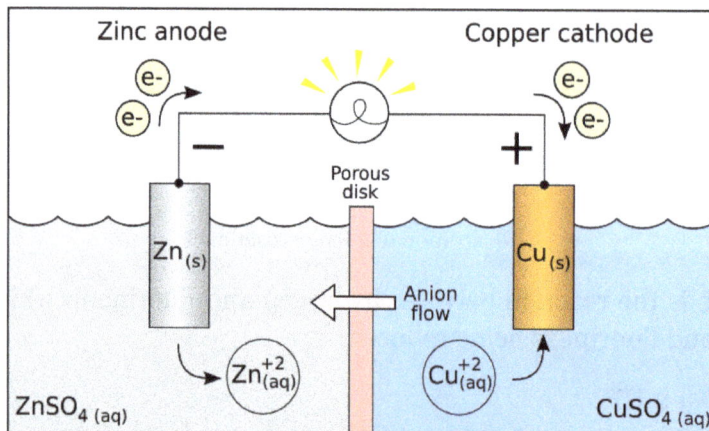

A redox reaction is the force behind an electrochemical cell like the Galvanic cell pictured. The battery is made out of a zinc electrode in a $ZnSO_4$ solution connected with a wire and a porous disk to a copper electrode in a $CuSO_4$ solution.

In this type of reaction, a metal atom in a compound (or in a solution) is replaced by an atom of another metal. For example, copper is deposited when zinc metal is placed in a copper(II) sulfate solution:

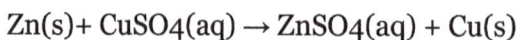

$$Zn(s) + CuSO_4(aq) \rightarrow ZnSO_4(aq) + Cu(s)$$

In the above reaction, zinc metal displaces the copper(II) ion from copper sulfate solution and thus liberates free copper metal.

The ionic equation for this reaction is:

$$Zn + Cu^{2+} \rightarrow Zn^{2+} + Cu$$

As two half-reactions, it is seen that the zinc is oxidized:

$$Zn \rightarrow Zn^{2+} + 2\ e^-$$

And the copper is reduced:

$$Cu^{2+} + 2\ e^- \rightarrow Cu$$

Other examples

- The reduction of nitrate to nitrogen in the presence of an acid (denitrification):

$$2\ NO{-}3 + 10\ e^- + 12\ H^+ \rightarrow N_2 + 6\ H_2O$$

- The combustion of hydrocarbons, such as in an internal combustion engine, which produces water, carbon dioxide, some partially oxidized forms such as carbon monoxide, and heat energy. Complete oxidation of materials containing carbon produces carbon dioxide.

- In organic chemistry, the stepwise oxidation of a hydrocarbon by oxygen produces water and, successively, an alcohol, an aldehyde or a ketone, a carboxylic acid, and then a peroxide.

Corrosion and Rusting

Oxides, such as iron(III) oxide or rust, which consists of hydrated iron(III) oxides $Fe_2O_3 \cdot nH_2O$ and iron(III) oxide-hydroxide (FeO(OH), Fe(OH)$_3$), form when oxygen combines with other elements

Iron rusting in pyrite cubes

- The term corrosion refers to the electrochemical oxidation of metals in reaction with an oxidant such as oxygen. Rusting, the formation of iron oxides, is a well-known example of electrochemical corrosion; it forms as a result of the oxidation of iron metal. Common rust often refers to iron(III) oxide, formed in the following chemical reaction:

$$4\ Fe + 3\ O_2 \rightarrow 2\ Fe_2O_3$$

- The oxidation of iron(II) to iron(III) by hydrogen peroxide in the presence of an acid:

$$Fe^{2+} \rightarrow Fe^{3+} + e^-$$

$$H_2O_2 + 2\ e^- \rightarrow 2\ OH^-$$

Overall equation:

$$2\ Fe^{2+} + H_2O_2 + 2\ H^+ \rightarrow 2\ Fe^{3+} + 2\ H_2O$$

Redox Reactions in Industry

Cathodic protection is a technique used to control the corrosion of a metal surface by making it the cathode of an electrochemical cell. A simple method of protection connects protected metal to a more easily corroded "sacrificial anode" to act as the anode. The sacrificial metal instead of the protected metal, then, corrodes. A common application of cathodic protection is in galvanized steel, in which a sacrificial coating of zinc on steel parts protects them from rust.

The primary process of reducing ore at high temperature to produce metals is known as smelting.

Oxidation is used in a wide variety of industries such as in the production of cleaning products and oxidizing ammonia to produce nitric acid, which is used in most fertilizers.

Redox reactions are the foundation of electrochemical cells, which can generate electrical energy or support electrosynthesis.

The process of electroplating uses redox reactions to coat objects with a thin layer of a material, as in chrome-plated automotive parts, silver plating cutlery, and gold-plated jewelry.

The production of compact discs depends on a redox reaction, which coats the disc with a thin layer of metal film.

Redox Reactions in Biology

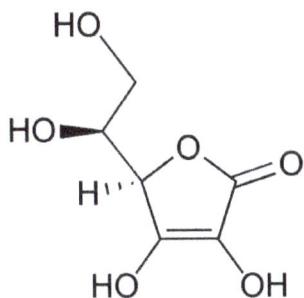

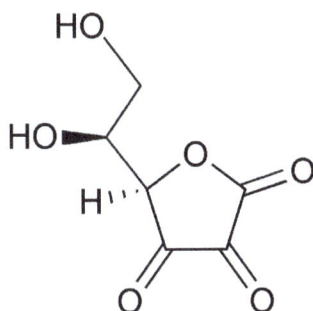

Top: ascorbic acid (reduced form of Vitamin C)
Bottom: dehydroascorbic acid (oxidized form of Vitamin C)

Many important biological processes involve redox reactions.

Cellular respiration, for instance, is the oxidation of glucose ($C_6H_{12}O_6$) to CO_2 and the reduction of oxygen to water. The summary equation for cell respiration is:

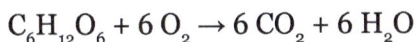

$$C_6H_{12}O_6 + 6\,O_2 \rightarrow 6\,CO_2 + 6\,H_2O$$

The process of cell respiration also depends heavily on the reduction of NAD^+ to NADH and the reverse reaction (the oxidation of NADH to NAD^+). Photosynthesis and cellular respiration are complementary, but photosynthesis is not the reverse of the redox reaction in cell respiration:

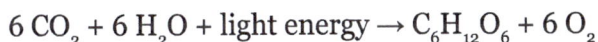

$$6\,CO_2 + 6\,H_2O + \text{light energy} \rightarrow C_6H_{12}O_6 + 6\,O_2$$

Biological energy is frequently stored and released by means of redox reactions. Photosynthesis involves the reduction of carbon dioxide into sugars and the oxidation of water into molecular oxygen. The reverse reaction, respiration, oxidizes sugars to produce carbon dioxide and water. As intermediate steps, the reduced carbon compounds are used to reduce nicotinamide adenine dinucleotide (NAD^+), which then contributes to the creation of a proton gradient, which drives the synthesis of adenosine triphosphate (ATP) and is maintained by the reduction of oxygen. In animal cells, mitochondria perform similar functions.

Free radical reactions are redox reactions that occur as a part of homeostasis and killing microorganisms, where an electron detaches from a molecule and then reattaches almost instantaneously. Free radicals are a part of redox molecules and can become harmful to the human body if they do not reattach to the redox molecule or an antioxidant. Unsatisfied free radicals can spur the mutation of cells they encounter and are, thus, causes of cancer.

The term redox state is often used to describe the balance of GSH/GSSG, NAD^+/NADH and $NADP^+$/NADPH in a biological system such as a cell or organ. The redox state is reflected in the balance of several sets of metabolites (e.g., lactate and pyruvate, beta-hydroxybutyrate, and acetoacetate), whose interconversion is dependent on these ratios. An abnormal redox state can develop in a variety of deleterious situations, such as hypoxia, shock, and sepsis. Redox mechanism also control some cellular processes. Redox proteins and their genes must be co-located for redox regulation according to the CoRR hypothesis for the function of DNA in mitochondria and chloroplasts.

Redox Cycling

A wide variety of aromatic compounds are enzymatically reduced to form free radicals that contain one more electron than their parent compounds. In general, the electron donor is any of a wide variety of flavoenzymes and their coenzymes. Once formed, these anion free radicals reduce molecular oxygen to superoxide, and regenerate the unchanged parent compound. The net reaction is the oxidation of the flavoenzyme's coenzymes and the reduction of molecular oxygen to form superoxide. This catalytic behavior has been described as futile cycle or redox cycling.

Examples of redox cycling-inducing molecules are the herbicide paraquat and other viologens and quinones such as menadione.

Redox Reactions in Geology

In geology, redox is important to both the formation of minerals and the mobilization of minerals, and is also important in some depositional environments. In general, the redox state of most rocks can be seen in the color of the rock. The rock forms in oxidizing conditions, giving it a red color. It is then "bleached" to a green—or sometimes white—form when a reducing fluid passes through the rock. The reduced fluid can also carry uranium-bearing minerals. Famous examples of redox conditions affecting geological processes include uranium deposits and Moqui marbles.

Mi Vida uranium mine, near Moab, Utah. The alternating red and white/green bands of sandstone correspond to oxidized and reduced conditions in groundwater redox chemistry.

Balancing Redox Reactions

Describing the overall electrochemical reaction for a redox process requires a *balancing* of the component half-reactions for oxidation and reduction. In general, for reactions in aqueous solution, this involves adding H^+, OH^-, H_2O, and electrons to compensate for the oxidation changes.

Acidic Media

In acidic media, H^+ ions and water are added to half-reactions to balance the overall reaction.

For instance, when manganese(II) reacts with sodium bismuthate:

Unbalanced reaction:	$Mn^{2+}(aq) + NaBiO_3(s) \rightarrow Bi^{3+}(aq) + MnO-4 (aq)$
Oxidation:	$4\ H_2O(l) + Mn^{2+}(aq) \rightarrow MnO-4(aq) + 8\ H^+(aq) + 5\ e^-$
Reduction:	$2\ e^- + 6\ H^+ + BiO-3(s) \rightarrow Bi^{3+}(aq) + 3\ H_2O(l)$

The reaction is balanced by scaling the two half-cell reactions to involve the same number of electrons (multiplying the oxidation reaction by the number of electrons in the reduction step and vice versa):

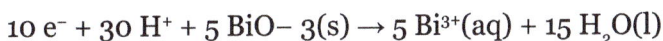

$$8\ H_2O(l) + 2\ Mn^{2+}(aq) \rightarrow 2\ MnO-4(aq) + 16\ H^+(aq) + 10\ e^-$$

$$10\ e^- + 30\ H^+ + 5\ BiO-3(s) \rightarrow 5\ Bi^{3+}(aq) + 15\ H_2O(l)$$

Adding these two reactions eliminates the electrons terms and yields the balanced reaction:

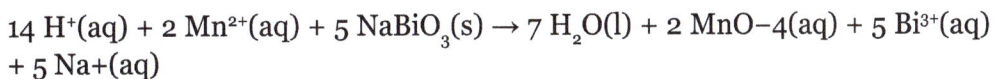

$$14\ H^+(aq) + 2\ Mn^{2+}(aq) + 5\ NaBiO_3(s) \rightarrow 7\ H_2O(l) + 2\ MnO-4(aq) + 5\ Bi^{3+}(aq) + 5\ Na+(aq)$$

Basic Media

In basic media, OH^- ions and water are added to half reactions to balance the overall reaction.

For example, in the reaction between potassium permanganate and sodium sulfite:

Unbalanced reaction:	$KMnO_4 + Na_2SO_3 + H_2O \rightarrow MnO_2 + Na_2SO_4 + KOH$
Reduction:	$3\ e^- + 2\ H_2O + MnO-4 \rightarrow MnO_2 + 4\ OH^-$
Oxidation:	$2\ OH^- + SO2- 3 \rightarrow SO2- 4 + H_2O + 2\ e^-$

Balancing the number of electrons in the two half-cell reactions gives:

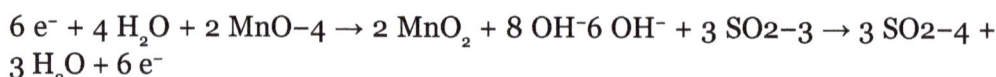

$$6\ e^- + 4\ H_2O + 2\ MnO-4 \rightarrow 2\ MnO_2 + 8\ OH^- 6\ OH^- + 3\ SO2-3 \rightarrow 3\ SO2-4 + 3\ H_2O + 6\ e^-$$

Adding these two half-cell reactions together gives the balanced equation:

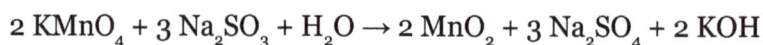

$$2\ KMnO_4 + 3\ Na_2SO_3 + H_2O \rightarrow 2\ MnO_2 + 3\ Na_2SO_4 + 2\ KOH$$

Memory Aids

The key terms involved in redox are often confusing to students. For example, an element that is oxidized loses electrons; however, that element is referred to as the reducing agent. Likewise, an element that is reduced gains electrons and is referred to as the oxidizing agent. Acronyms or mnemonics are commonly used to help remember the terminology:

- "OIL RIG" — oxidation is loss of electrons, reduction is gain of electrons.

- "LEO the lion says GER" — loss of electrons is oxidation, gain of electrons is reduction.

- "LEORA says GEROA" — loss of electrons is oxidation (reducing agent), gain of electrons is reduction (oxidizing agent).

- "RED CAT" and "AN OX", or "AnOx RedCat" ("an ox-red cat") — reduction occurs at the cathode and the anode is for oxidation.

- "RED CAT gains what AN OX loses" – reduction at the cathode gains (electrons) what anode oxidation loses (electrons).

Alkylation

Alkylation is the transfer of an alkyl group from one molecule to another. The alkyl group may be transferred as an alkyl carbocation, a free radical, a carbanion or a carbene (or their equivalents). An alkyl group is a piece of a molecule with the general formula C_nH_{2n+1}, where n is the integer depicting the number of carbons linked together.

For example, a methyl group ($n = 1$, CH_3) is a fragment of a methane molecule (CH_4). Alkylating agents utilize selective alkylation by adding the desired aliphatic carbon chain to the previously chosen starting molecule. This is one of many known chemical syntheses. Alkyl groups can also be removed in a process known as dealkylation.

In oil refining contexts, alkylation refers to a particular alkylation of isobutane with olefins. For upgrading of petroleum, alkylation produces synthetic C_7–C_8 alkylate, which is a premium blending stock for gasoline.

In medicine, alkylation of DNA is used in chemotherapy to damage the DNA of cancer cells. Alkylation is accomplished with the class of drugs called alkylating antineoplastic agents.

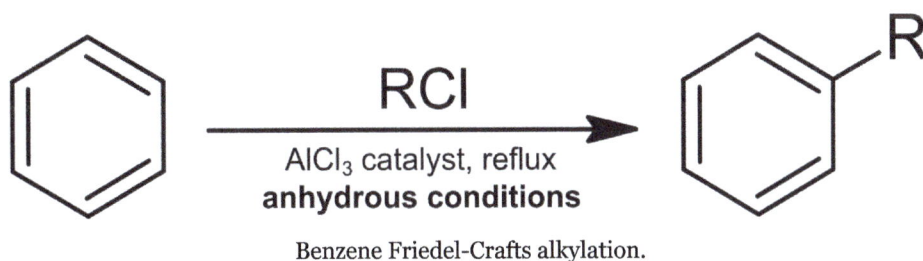

Benzene Friedel-Crafts alkylation.

Alkylating Agents

Alkylating agents are classified according to their nucleophilic or electrophilic character.

Nucleophilic Alkylating Agents

Nucleophilic alkylating agents deliver the equivalent of an alkyl anion (carbanion). Examples include the use of organometallic compounds such as Grignard (organomagnesium), organolithium, organocopper, and organosodium reagents. These compounds typically can add to an electron-deficient carbon atom such as at a carbonyl group. Nucleophilic alkylating agents can also displace halide substituents on a carbon atom. In the presence of catalysts, they also alkylate alkyl and aryl halides, as exemplified by Suzuki couplings.

Electrophilic Alkylating Agents

Electrophilic alkylating agents deliver the equivalent of an alkyl cation. Examples include the use of alkyl halides with a Lewis acid catalyst to alkylate aromatic substrates in Friedel-Crafts reactions. Alkyl halides can also react directly with amines to form C-N bonds; the same holds true for other nucleophiles such as alcohols, carboxylic acids, thiols, etc. Trimethyloxonium tetrafluoroborate and triethyloxonium tetrafluoroborate are particularly strong electrophiles due to their overt positive charge and an inert leaving group (dimethyl or diethyl ether).

Electrophilic, soluble alkylating agents are often very toxic, due to their ability to alkylate DNA. They should be handled with proper PPE. This mechanism of toxicity is also responsible for the ability of some alkylating agents to perform as anti-cancer drugs in the form of alkylating antineoplastic agents, and also as chemical weapons such as mustard gas. Alkylated DNA either does not coil or uncoil properly, or cannot be processed by information-decoding enzymes. This results in cytotoxicity with the effects of inhibition the growth of the cell, initiation of programmed cell death or apoptosis. However, mutations are also triggered, including carcinogenic mutations, explaining the higher incidence of cancer after exposure.

Alcohols and phenols can be alkylated to give alkyl ethers:

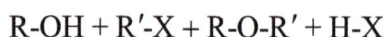

$$R\text{-}OH + R'\text{-}X + R\text{-}O\text{-}R' + H\text{-}X$$

The produced acid HX is neutralized with a base, or, alternatively, the alcohol is deprotonated first to give an alkoxide or phenoxide. For example, dimethyl sulfate alkylates the sodium salt of phenol to give anisole, the methyl ether of phenol. The dimethyl sulfate is dealkylated to sodium methylsulfate.

$$Ph\text{-}O^- + Me_2\text{-}SO_4 + Ph\text{-}O\text{-}Me + Me\text{-}SO_4^- \text{ (with } Na^+ \text{ as a spectator ion)}$$

On the contrary, the alkylation of amines introduces the problem that the alkylation of an amine makes it *more* nucleophilic. Thus, when an electrophilic alkylating agent is introduced to a primary amine, it will preferentially alkylate all the way to a quaternary ammonium cation.

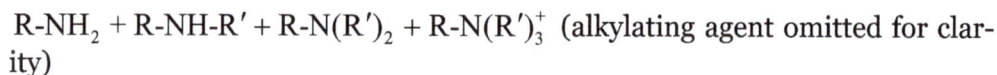

$$R\text{-}NH_2 + R\text{-}NH\text{-}R' + R\text{-}N(R')_2 + R\text{-}N(R')_3^+ \text{ (alkylating agent omitted for clarity)}$$

If the quaternary ammonium is not the desired product, more circuitous routes such as reductive amination are necessary.

Carbene Alkylating Agents

Carbenes are extremely reactive and are known to attack even unactivated C-H bonds. Carbenes can be generated by elimination of a diazo group. Unlike electrophilic or nucleophilic alkylating agents, carbenes are neutral, and they insert into bonds rather than discard leaving groups. A metal can form a carbene equivalent called a transition metal carbene complex.

In Biology

Methylation is the most common type of alkylation, being associated with the transfer of a methyl group. Methylation is distinct from alkylation in that it is specifically the transfer of one carbon, whereas alkylation can refer to the transfer of long chain carbon groups. Methylation in nature is typically effected by vitamin B12-derived enzymes, where the methyl group is carried by cobalt. In methanogenesis, coenzyme M is methylated by tetrahydromethanopterin.

Electrophilic compounds may alkylate different nucleophiles in the body. The toxicity, carcinogenity, and paradoxically, cancer cell-killing abilities of different DNA alkylating agents are an example.

Demethylation is the reverse of methylation.

Oil Refining

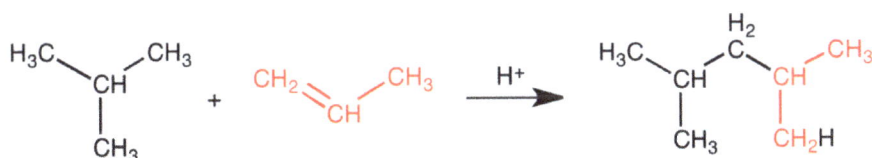

Alkylation of alkenes (shown in red is propene) by isobutane is a major process in refineries to produce higher octane "alkylate" for gasoline blending, in this example yielding isoheptane. It is catalysed by strong acids such as hydrofluoric acid (HF) and sulfuric acid (H2SO4).

In a standard oil refinery process, isobutane is alkylated with low-molecular-weight alkenes (primarily a mixture of propene and butene) in the presence of a Bronsted acid catalyst, either sulfuric acid or hydrofluoric acid. In an oil refinery it is referred to as a sulfuric acid alkylation unit (SAAU) or a hydrofluoric alkylation unit, (HFAU). Refinery workers may simply refer to it as the alky or alky unit. The catalyst protonates the alkenes (propene, butene) to produce reactive carbocations, which alkylate isobutane. The reaction is carried out at mild temperatures (0 and 30 °C) in a two-phase reaction. Because the reaction is exothermic, cooling is needed: SAAU plants require lower temperatures so the cooling medium needs to be chilled, for HFAU normal refinery cooling water will suffice. It is important to keep a high ratio of isobutane to alkene at the point of reaction to prevent side reactions which produces a lower octane product, so the plants have a high recycle of isobutane back to feed. The phases separate spontaneously, so the acid phase is vigorously mixed with the hydrocarbon phase to create sufficient contact surface.

The product is called alkylate and is composed of a mixture of high-octane, branched-chain paraffinic hydrocarbons (mostly isoheptane and isooctane). Alkylate is a premium gasoline blending stock because it has exceptional antiknock properties and is clean burning. Alkylate is also a key component of avgas. The octane number of the alkylate depends mainly upon the kind of alkenes used and upon operating conditions. For example, isooctane results from combining butylene with isobutane and has an octane rating of 100 by definition. There are other products in the alkylate, so the octane rating will vary accordingly.

Since crude oil generally contains only 10 to 40 percent of hydrocarbon constituents in the gasoline range, refineries use a fluid catalytic cracking process to convert high molecular weight hydrocarbons into smaller and more volatile compounds, which are then converted into liquid gasoline-size hydrocarbons. Alkylation processes transform low molecular-weight alkenes and iso-paraffin molecules into larger iso-paraffins with a high octane number.

Combining cracking, polymerization, and alkylation can result in a gasoline yield representing 70 percent of the starting crude oil. More advanced processes, such as cyclicization of paraffins and dehydrogenation of naphthenes forming aromatic hydrocarbons in a catalytic reformer, have also been developed to increase the octane rating of gasoline. Modern refinery operation can be shifted to produce almost any fuel type with specified performance criteria from a single crude feedstock.

Refineries examine whether it makes sense economically to install alkylation units. Alkylation units are complex, with substantial economy of scale. In addition to a suitable quantity of feedstock, the price spread between the value of alkylate product and alternate feedstock disposition value must be large enough to justify the installation. Alternative outlets for refinery alklylation feedstocks include sales as LPG, blending of C4 streams directly into gasoline to lower the flash point of the product and feedstocks for chemical plants. Local market conditions vary widely between plants. Variation in the RVP (Reid vapor pressure) specification for gasoline between countries and between seasons dramatically impacts the amount of butane streams that can be blended directly into gasoline. The transportation of specific types of LPG streams can be expensive so local disparities in economic conditions are often not fully mitigated by cross market movements of alkylation feedstocks.

The availability of a suitable catalyst is also an important factor in deciding whether to build an alkylation plant. If sulfuric acid is used, significant volumes are needed. Access to a suitable plant is required for the supply of fresh acid and the disposition of spent acid. If a sulfuric acid plant must be constructed specifically to support an alkylation unit, such construction will have a significant impact on both the initial requirements for capital and ongoing costs of operation. Alternatively it is possible to install a WSA Process unit to regenerate the spent acid. No drying of the gas takes place. This means that there will be no loss of acid, no acidic waste material and no heat is lost in process gas reheating. The selective condensation in the WSA condenser ensures that the regenerated fresh acid will be 98% w/w even with the humid process gas. It is possible to combine spent acid regeneration with disposal of hydrogen sulfide by using the hydrogen sulfide as internal fuel in the refinery or elsewhere.

The second main catalyst option is hydrofluoric acid. In typical alkylation plants, rates of consumption for acid are much lower than for sulfuric acid. These plants also produce alkylate with better octane rating than do sulfuric plants. However, due to its hazardous nature, HF acid is produced at very few locations and transportation must be managed rigorously.

Chemical Reaction

A chemical reaction is a process that leads to the transformation of one set of chemical substances to another. Classically, chemical reactions encompass changes that only involve the positions of electrons in the forming and breaking of chemical bonds between atoms, with no change to the nuclei (no change to the elements present), and can often be described by a chemical equation. Nuclear chemistry is a sub-discipline of chemistry that involves the chemical reactions of unstable and radioactive elements where both electronic and nuclear changes may occur.

A thermite reaction using iron(III) oxide. The sparks flying outwards are globules of molten iron trailing smoke in their wake.

The substance (or substances) initially involved in a chemical reaction are called reactants or reagents. Chemical reactions are usually characterized by a chemical change, and they yield one or more products, which usually have properties different from the reactants. Reactions often consist of a sequence of individual sub-steps, the so-called elementary reactions, and the information on the precise course of action is part of the reaction mechanism. Chemical reactions are described with chemical equations, which symbolically present the starting materials, end products, and sometimes intermediate products and reaction conditions.

Chemical reactions happen at a characteristic reaction rate at a given temperature and chemical concentration. Typically, reaction rates increase with increasing temperature because there is more thermal energy available to reach the activation energy necessary for breaking bonds between atoms.

Reactions may proceed in the forward or reverse direction until they go to completion or reach equilibrium. Reactions that proceed in the forward direction to approach equilibrium are often described as spontaneous, requiring no input of free energy to go forward. Non-spontaneous reactions require input of free energy to go forward (examples include charging a battery by applying an external electrical power source, or photosynthesis driven by absorption of electromagnetic radiation in the form of sunlight).

Different chemical reactions are used in combinations during chemical synthesis in order to obtain a desired product. In biochemistry, a consecutive series of chemical reactions (where the product of one reaction is the reactant of the next reaction) form metabolic pathways. These reactions are often catalyzed by protein enzymes. Enzymes increase the rates of biochemical reactions, so that metabolic syntheses and decompositions impossible under ordinary conditions can occur at the temperatures and concentrations present within a cell.

The general concept of a chemical reaction has been extended to reactions between entities smaller than atoms, including nuclear reactions, radioactive decays, and reactions between elementary particles as described by quantum field theory.

History

Chemical reactions such as combustion in fire, fermentation and the reduction of ores to metals were known since antiquity. Initial theories of transformation of materials were developed by Greek philosophers, such as the Four-Element Theory of Empedocles stating that any substance is composed of the four basic elements – fire, water, air and earth. In the Middle Ages, chemical transformations were studied by Alchemists. They attempted, in particular, to convert lead into gold, for which purpose they used reactions of lead and lead-copper alloys with sulfur.

Antoine Lavoisier developed the theory of combustion as a chemical reaction with oxygen

The production of chemical substances that do not normally occur in nature has long been tried, such as the synthesis of sulfuric and nitric acids attributed to the controversial alchemist Jābir ibn Hayyān. The process involved heating of sulfate and nitrate minerals such as copper sulfate, alum and saltpeter. In the 17th century, Johann Rudolph Glauber produced hydrochloric acid and sodium sulfate by reacting sulfuric acid and sodium chloride. With the development of the lead chamber process in 1746 and the Leblanc process, allowing large-scale production of sulfuric acid and sodium carbonate, respectively, chemical reactions became implemented into the industry. Further optimization of sulfuric acid technology resulted in the contact process in the 1880s, and the Haber process was developed in 1909–1910 for ammonia synthesis.

From the 16th century, researchers including Jan Baptist van Helmont, Robert Boyle and Isaac Newton tried to establish theories of the experimentally observed chemical transformations. The phlogiston theory was proposed in 1667 by Johann Joachim Becher. It postulated the existence of a fire-like element called "phlogiston", which was contained within combustible bodies and released during combustion. This proved to be false in 1785 by Antoine Lavoisier who found the correct explanation of the combustion as reaction with oxygen from the air.

Joseph Louis Gay-Lussac recognized in 1808 that gases always react in a certain relationship with each other. Based on this idea and the atomic theory of John Dalton, Joseph Proust had developed the law of definite proportions, which later resulted in the concepts of stoichiometry and chemical equations.

Regarding the organic chemistry, it was long believed that compounds obtained from living organisms were too complex to be obtained synthetically. According to the concept of vitalism, organic matter was endowed with a "vital force" and distinguished from inorganic materials. This separation was ended however by the synthesis of urea from inorganic precursors by Friedrich Wöhler in 1828. Other chemists who brought major contributions to organic chemistry include Alexander William Williamson with his synthesis of ethers and Christopher Kelk Ingold, who, among many discoveries, established the mechanisms of substitution reactions.

Equations

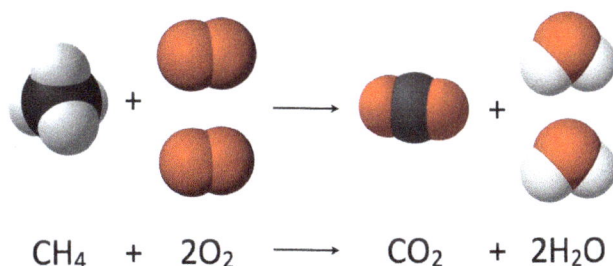

$$CH_4 \; + \; 2O_2 \; \longrightarrow \; CO_2 \; + \; 2H_2O$$

As seen from the equation CH4 + 2 O2 → CO2 + 2 H2O, a coefficient of 2 must be placed before the oxygen gas on the reactants side and before the water on the products side in order for, as per the law of conservation of mass, the quantity of each element does not change during the reaction

Chemical equations are used to graphically illustrate chemical reactions. They consist of chemical or structural formulas of the reactants on the left and those of the products on the right. They are separated by an arrow ($\rightarrow$) which indicates the direction and type of the reaction; the arrow is read as the word "yields". The tip of the arrow points in the direction in which the reaction proceeds. A double arrow ($\rightleftharpoons$) pointing in opposite directions is used for equilibrium reactions. Equations should be balanced according to the stoichiometry, the number of atoms of each species should be the same on both sides of the equation. This is achieved by scaling the number of involved molecules (A, B, C and D in a schematic example below) by the appropriate integers a, b, c and d.

$$aA + bB -> cC + dD$$

More elaborate reactions are represented by reaction schemes, which in addition to starting materials and products show important intermediates or transition states. Also, some relatively minor additions to the reaction can be indicated above the reaction arrow; examples of such additions are water, heat, illumination, a catalyst, etc. Similarly, some minor products can be placed below the arrow, often with a minus sign.

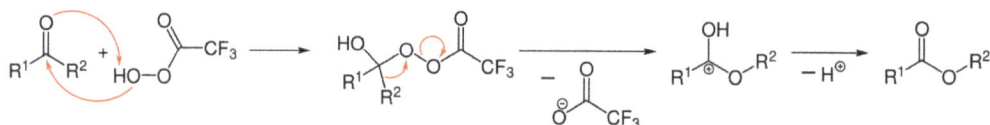

An example of organic reaction: oxidation of ketones to esters with a peroxycarboxylic acid

Retrosynthetic analysis can be applied to design a complex synthesis reaction. Here the analysis starts from the products, for example by splitting selected chemical bonds, to arrive at plausible initial reagents. A special arrow ($\Rightarrow$) is used in retro reactions.

Elementary Reactions

The elementary reaction is the smallest division into which a chemical reaction can be decomposed, it has no intermediate products. Most experimentally observed reactions are built up from many elementary reactions that occur in parallel or sequentially. The actual sequence of the individual elementary reactions is known as reaction mechanism. An elementary reaction involves a few molecules, usually one or two, because of the low probability for several molecules to meet at a certain time.

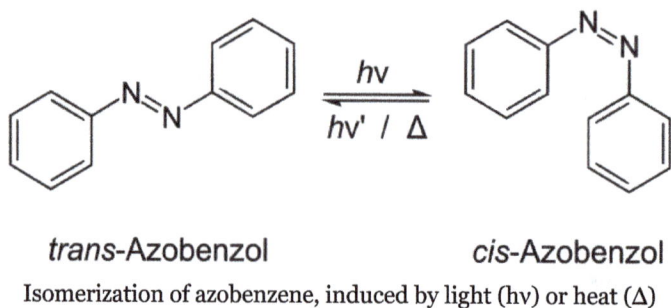

trans-Azobenzol   *cis*-Azobenzol

Isomerization of azobenzene, induced by light (hv) or heat (Δ)

The most important elementary reactions are unimolecular and bimolecular reactions. Only one molecule is involved in a unimolecular reaction; it is transformed by an isomerization or a dissociation into one or more other molecules. Such reactions require the addition of energy in the form of heat or light. A typical example of a unimolecular reaction is the cis–trans isomerization, in which the cis-form of a compound converts to the trans-form or vice versa.

In a typical dissociation reaction, a bond in a molecule splits (ruptures) resulting in two molecular fragments. The splitting can be homolytic or heterolytic. In the first case, the bond is divided so that each product retains an electron and becomes a neutral radical. In the second case, both electrons of the chemical bond remain with one of the products, resulting in charged ions. Dissociation plays an important role in triggering chain reactions, such as hydrogen–oxygen or polymerization reactions.

$$AB -> A + B$$

Dissociation of a molecule AB into fragments A and B

For bimolecular reactions, two molecules collide and react with each other. Their merger is called chemical synthesis or an addition reaction.

$$A + B -> AB$$

Another possibility is that only a portion of one molecule is transferred to the other molecule. This type of reaction occurs, for example, in redox and acid-base reactions. In redox reactions, the transferred particle is an electron, whereas in acid-base reactions it is a proton. This type of reaction is also called metathesis.

$$HA + B -> A + HB$$

for example

$$NaCl + AgNO3 -> NaNO3 + AgCl(v)$$

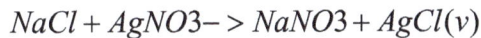

Chemical Equilibrium

Most chemical reactions are reversible, that is they can and do run in both directions. The forward and reverse reactions are competing with each other and differ in reaction rates. These rates depend on the concentration and therefore change with time of the reaction: the reverse rate gradually increases and becomes equal to the rate of the forward reaction, establishing the so-called chemical equilibrium. The time to reach equilibrium depends on such parameters as temperature, pressure and the materials involved, and is determined by the minimum free energy. In equilibrium, the Gibbs free energy must be zero. The pressure dependence can be explained with the Le Chatelier's principle. For example, an increase in pressure due to decreasing volume causes the reaction to shift to the side with the fewer moles of gas.

The reaction yield stabilizes at equilibrium, but can be increased by removing the product from the reaction mixture or changed by increasing the temperature or pressure. A change in the concentrations of the reactants does not affect the equilibrium constant, but does affect the equilibrium position.

Thermodynamics

Chemical reactions are determined by the laws of thermodynamics. Reactions can proceed by themselves if they are exergonic, that is if they release energy. The associated free energy of the reaction is composed of two different thermodynamic quantities, enthalpy and entropy:

$$\Delta G = \Delta H - T \cdot \Delta S.$$

G: free energy, H: enthalpy, T: temperature, S: entropy, Δ: difference(change between original and product)

Reactions can be exothermic, where ΔH is negative and energy is released. Typical examples of exothermic reactions are precipitation and crystallization, in which ordered solids are formed from disordered gaseous or liquid phases. In contrast, in endothermic reactions, heat is consumed from the environment. This can occur by increasing the entropy of the system, often through the formation of gaseous reaction products, which have high entropy. Since the entropy increases with temperature, many endothermic reactions preferably take place at high temperatures. On the contrary, many exothermic reactions such as crystallization occur at low temperatures. Changes in temperature can sometimes reverse the sign of the enthalpy of a reaction, as for the carbon monoxide reduction of molybdenum dioxide:

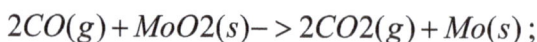

$$2CO(g) + MoO2(s) -> 2CO2(g) + Mo(s);$$

$$\Delta H^\circ = +21.86 \text{ kJ at 298 K}$$

This reaction to form carbon dioxide and molybdenum is endothermic at low temperatures, becoming less so with increasing temperature. ΔH° is zero at 1855 K, and the reaction becomes exothermic above that temperature.

Changes in temperature can also reverse the direction tendency of a reaction. For example, the water gas shift reaction

$$CO(g) + H2O(v) <=> CO2(g) + H2(g)$$

is favored by low temperatures, but its reverse is favored by high temperature. The shift in reaction direction tendency occurs at 1100 K.

Reactions can also be characterized by the internal energy which takes into account changes in the entropy, volume and chemical potential. The latter depends, among other things, on the activities of the involved substances.

$$dU = T \cdot dS - p \cdot dV + \mu \cdot dn$$

U: internal energy, S: entropy, p: pressure, μ: chemical potential, n: number of molecules, d: small change sign

Kinetics

The speed at which reactions takes place is studied by reaction kinetics. The rate depends on various parameters, such as:

- Reactant concentrations, which usually make the reaction happen at a faster rate if raised through increased collisions per unit time. Some reactions, however, have rates that are *independent* of reactant concentrations. These are called zero order reactions.

- Surface area available for contact between the reactants, in particular solid ones in heterogeneous systems. Larger surface areas lead to higher reaction rates.

- Pressure – increasing the pressure decreases the volume between molecules and therefore increases the frequency of collisions between the molecules.

- Activation energy, which is defined as the amount of energy required to make the reaction start and carry on spontaneously. Higher activation energy implies that the reactants need more energy to start than a reaction with a lower activation energy.

- Temperature, which hastens reactions if raised, since higher temperature increases the energy of the molecules, creating more collisions per unit time,

- The presence or absence of a catalyst. Catalysts are substances which change the pathway (mechanism) of a reaction which in turn increases the speed of a reaction by lowering the activation energy needed for the reaction to take place. A catalyst is not destroyed or changed during a reaction, so it can be used again.

- For some reactions, the presence of electromagnetic radiation, most notably ultraviolet light, is needed to promote the breaking of bonds to start the reaction. This is particularly true for reactions involving radicals.

Several theories allow calculating the reaction rates at the molecular level. This field is referred to as reaction dynamics. The rate v of a first-order reaction, which could be disintegration of a substance A, is given by:

$$v = -\frac{d[A]}{dt} = k \cdot [A].$$

Its integration yields:

$$[A](t) = [A]_0 \cdot e^{-k \cdot t}.$$

Here k is first-order rate constant having dimension 1/time, [A](t) is concentration at a time t and [A]$_0$ is the initial concentration. The rate of a first-order reaction depends only on the concentration and the properties of the involved substance, and the reaction itself can be described with the characteristic half-life. More than one time constant is needed when describing reactions of higher order. The temperature dependence of the rate constant usually follows the Arrhenius equation:

$$k = k_0 e - E_a >$$

where E$_a$ is the activation energy and k$_B$ is the Boltzmann constant. One of the simplest models of reaction rate is the collision theory. More realistic models are tailored to a specific problem and include the transition state theory, the calculation of the potential energy surface, the Marcus theory and the Rice–Ramsperger–Kassel–Marcus (RRKM) theory.

Reaction Types

Four Basic Types

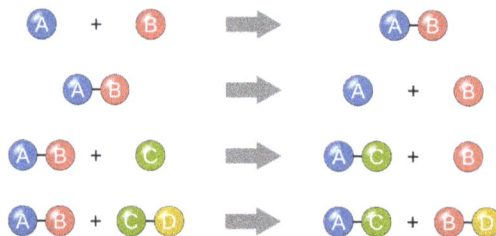

Representation of four basic chemical reactions types: synthesis, decomposition, single replacement and double replacement.

Synthesis

In a synthesis reaction, two or more simple substances combine to form a more complex substance. These reactions are in the general form:

$$A + B - > AB$$

Two or more reactants yielding one product is another way to identify a synthesis reaction. One example of a synthesis reaction is the combination of iron and sulfur to form iron(II) sulfide:

$$8Fe + S8 - > 8FeS$$

Another example is simple hydrogen gas combined with simple oxygen gas to produce a more complex substance, such as water.

Decomposition

A decomposition reaction is when a more complex substance breaks down into its more simple parts. It is thus the opposite of a synthesis reaction, and can be written as

$$2H2O -> 2H2 + O2$$

One example of a decomposition reaction is the electrolysis of water to make oxygen and hydrogen gas:

$$A + BC -> AC + B$$

Single Replacement

In a single replacement reaction, a single uncombined element replaces another in a compound; in other words, one element trades places with another element in a compound These reactions come in the general form of:

$$Mg + 2H2O -> Mg(OH)2 + H2 \uparrow$$

One example of a single displacement reaction is when magnesium replaces hydrogen in water to make magnesium hydroxide and hydrogen gas:

$$AB + CD -> AD + CB$$

Double Replacement

In a double replacement reaction, the anions and cations of two compounds switch places and form two entirely different compounds. These reactions are in the general form:

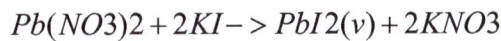

$$Pb(NO3)2 + 2KI -> PbI2(v) + 2KNO3$$

For example, when barium chloride ($BaCl_2$) and magnesium sulfate ($MgSO_4$) react, the SO_4^{2-} anion switches places with the $2Cl^-$ anion, giving the compounds $BaSO_4$ and $MgCl_2$.

Another example of a double displacement reaction is the reaction of lead(II) nitrate with potassium iodide to form lead(II) iodide and potassium nitrate:

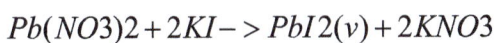

$$Pb(NO3)2 + 2KI -> PbI2(v) + 2KNO3$$

Oxidation and Reduction

Redox reactions can be understood in terms of transfer of electrons from one involved species (reducing agent) to another (oxidizing agent). In this process, the former species is *oxidized* and the latter is *reduced*. Though sufficient for many purposes, these descriptions are not precisely correct. Oxidation is better defined as an increase in oxidation state, and reduction as a decrease in oxidation state. In practice, the transfer of electrons will always change the oxidation state, but there are many reactions that are classed as "redox" even though no electron transfer occurs (such as those involving covalent bonds).

Sodium chloride is formed through the redox reaction of sodium metal and chlorine gas

In the following redox reaction, hazardous sodium metal reacts with toxic chlorine gas to form the ionic compound sodium chloride, or common table salt:

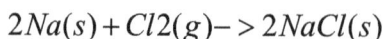

$$2Na(s) + Cl2(g) -> 2NaCl(s)$$

In the reaction, sodium metal goes from an oxidation state of 0 (as it is a pure element) to +1: in other words, the sodium lost one electron and is said to have been oxidized. On the other hand, the chlorine gas goes from an oxidation of 0 (it is also a pure element) to −1: the chlorine gains one electron and is said to have been reduced. Because the chlorine is the one reduced, it is considered the electron acceptor, or in other words, induces oxidation in the sodium – thus the chlorine gas is considered the oxidizing agent. Conversely, the sodium is oxidized or is the electron donor, and thus induces reduction in the other species and is considered the *reducing agent*.

Which of the involved reactants would be reducing or oxidizing agent can be predicted from the electronegativity of their elements. Elements with low electronegativity, such as most metals, easily donate electrons and oxidize – they are reducing agents. On the contrary, many ions with high oxidation numbers, such as $H2O2$, $MnO-4$, $CrO3$, $Cr2O2-7$, $OsO4$ can gain one or two extra electrons and are strong oxidizing agents.

The number of electrons donated or accepted in a redox reaction can be predicted from the electron configuration of the reactant element. Elements try to reach the low-energy noble gas configuration, and therefore alkali metals and halogens will donate and accept one electron respectively. Noble gases themselves are chemically inactive.

An important class of redox reactions are the electrochemical reactions, where electrons from the power supply are used as the reducing agent. These reactions are particularly important for the production of chemical elements, such as chlorine or aluminium. The reverse process in which electrons are released in redox reactions and can be used as electrical energy is possible and used in batteries.

Complexation

In complexation reactions, several ligands react with a metal atom to form a coordination complex. This is achieved by providing lone pairs of the ligand into empty orbitals of the metal atom and forming dipolar bonds. The ligands are Lewis bases, they can be both ions and neutral molecules, such as carbon monoxide, ammonia or water. The number of ligands that react with a central metal atom can be found using the 18-electron rule, saying that the valence shells of a transition metal will collectively accommodate 18 electrons, whereas the symmetry of the resulting complex can be predicted with the crystal field theory and ligand field theory. Complexation reactions also include ligand exchange, in which one or more ligands are replaced by another, and redox processes which change the oxidation state of the central metal atom.

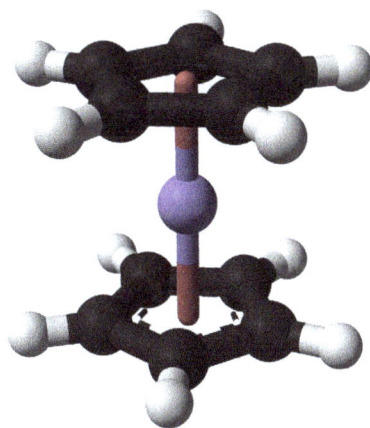

Ferrocene – an iron atom sandwiched between two C_5H_5 ligands

Acid-base Reactions

In the Brønsted–Lowry acid–base theory, an acid-base reaction involves a transfer of protons (H^+) from one species (the acid) to another (the base). When a proton is removed from an acid, the resulting species is termed that acid's conjugate base. When the proton is accepted by a base, the resulting species is termed that base's conjugate acid. In other words, acids act as proton donors and bases act as proton acceptors according to the following equation:

$$\underset{acid}{\underline{HA}} + \underset{base}{\underline{B}} <=> \underset{conjugated\ base}{\underline{A^-}} + \underset{conjugated\ acid}{\underline{HB+}}$$

The reverse reaction is possible, and thus the acid/base and conjugated base/acid are always in equilibrium. The equilibrium is determined by the acid and base dissociation constants (K_a and K_b) of the involved substances. A special case of the acid-base reaction is the neutralization where an acid and a base, taken at exactly same amounts, form a neutral salt.

Acid-base reactions can have different definitions depending on the acid-base concept employed. Some of the most common are:

- Arrhenius definition: Acids dissociate in water releasing H_3O^+ ions; bases dissociate in water releasing OH^- ions.

- Brønsted-Lowry definition: Acids are proton (H^+) donors, bases are proton acceptors; this includes the Arrhenius definition.

- Lewis definition: Acids are electron-pair acceptors, bases are electron-pair donors; this includes the Brønsted-Lowry definition.

Precipitation

Precipitation is the formation of a solid in a solution or inside another solid during a chemical reaction. It usually takes place when the concentration of dissolved ions exceeds the solubility limit and forms an insoluble salt. This process can be assisted by adding a precipitating agent or by removal of the solvent. Rapid precipitation results in an amorphous or microcrystalline residue and slow process can yield single crystals. The latter can also be obtained by recrystallization from microcrystalline salts.

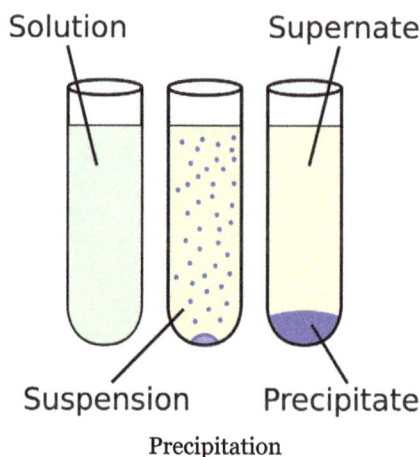

Precipitation

Solid-state Reactions

Reactions can take place between two solids. However, because of the relatively small diffusion rates in solids, the corresponding chemical reactions are very slow in comparison to liquid and gas phase reactions. They are accelerated by increasing the reaction temperature and finely dividing the reactant to increase the contacting surface area.

Reactions at the Solid|Gas Interface

Reaction can take place at the solid|gas interface, surfaces at very low pressure such as ultra-high vacuum. Via scanning tunneling microscopy, it is possible to observe reactions at the solid|gas interface in real space, if the time scale of the reaction is in the correct range. Reactions at the solid|gas interface are in some cases related to catalysis.

Photochemical Reactions

In photochemical reactions, atoms and molecules absorb energy (photons) of the illumination light and convert into an excited state. They can then release this energy by breaking chemical bonds, thereby producing radicals. Photochemical reactions include hydrogen–oxygen reactions, radical polymerization, chain reactions and rearrangement reactions.

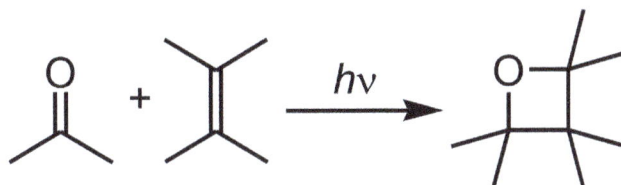

In this Paterno–Büchi reaction, a photoexcited carbonyl group is added to an unexcited olefin, yielding an oxetane.

Many important processes involve photochemistry. The premier example is photosynthesis, in which most plants use solar energy to convert carbon dioxide and water into glucose, disposing of oxygen as a side-product. Humans rely on photochemistry for the formation of vitamin D, and vision is initiated by a photochemical reaction of rhodopsin. In fireflies, an enzyme in the abdomen catalyzes a reaction that results in bioluminescence. Many significant photochemical reactions, such as ozone formation, occur in the Earth atmosphere and constitute atmospheric chemistry.

Catalysis

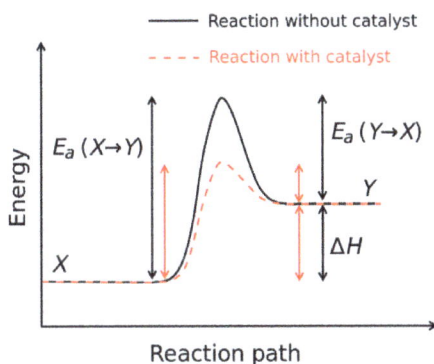

Schematic potential energy diagram showing the effect of a catalyst in an endothermic chemical reaction. The presence of a catalyst opens a different reaction pathway (in red) with a lower activation energy. The final result and the overall thermodynamics are the same.

Solid heterogeneous catalysts are plated on meshes in ceramic catalytic converters in order to maximize their surface area. This exhaust converter is from a Peugeot 106 S2 1100

In catalysis, the reaction does not proceed directly, but through reaction with a third substance known as catalyst. Although the catalyst takes part in the reaction, it is returned to its original state by the end of the reaction and so is not consumed. However, it can be inhibited, deactivated or destroyed by secondary processes. Catalysts can be used in a different phase (heterogeneous) or in the same phase (homogeneous) as the reactants. In heterogeneous catalysis, typical secondary processes include coking where the catalyst becomes covered by polymeric side products. Additionally, heterogeneous catalysts can dissolve into the solution in a solid–liquid system or evaporate in a solid–gas system. Catalysts can only speed up the reaction – chemicals that slow down the reaction are called inhibitors. Substances that increase the activity of catalysts are called promoters, and substances that deactivate catalysts are called catalytic poisons. With a catalyst, a reaction which is kinetically inhibited by a high activation energy can take place in circumvention of this activation energy.

Heterogeneous catalysts are usually solids, powdered in order to maximize their surface area. Of particular importance in heterogeneous catalysis are the platinum group metals and other transition metals, which are used in hydrogenations, catalytic reforming and in the synthesis of commodity chemicals such as nitric acid and ammonia. Acids are an example of a homogeneous catalyst, they increase the nucleophilicity of carbonyls, allowing a reaction that would not otherwise proceed with electrophiles. The advantage of homogeneous catalysts is the ease of mixing them with the reactants, but they may also be difficult to separate from the products. Therefore, heterogeneous catalysts are preferred in many industrial processes.

Reactions in Organic Chemistry

In organic chemistry, in addition to oxidation, reduction or acid-base reactions, a number of other reactions can take place which involve covalent bonds between carbon atoms or carbon and heteroatoms (such as oxygen, nitrogen, halogens, etc.). Many specific reactions in organic chemistry are name reactions designated after their discoverers.

Substitution

In a substitution reaction, a functional group in a particular chemical compound is replaced by another group. These reactions can be distinguished by the type of substituting species into a nucleophilic, electrophilic or radical substitution.

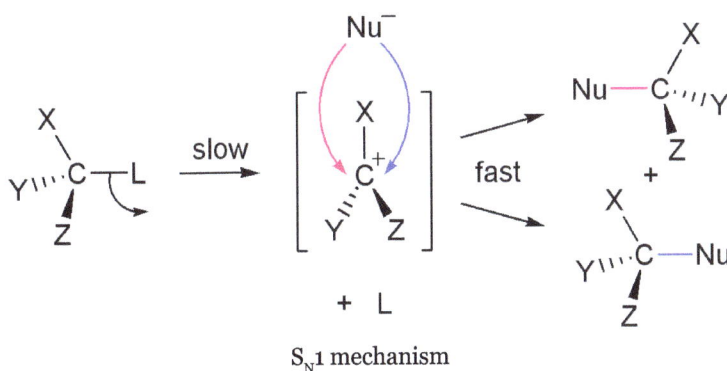

S_N1 mechanism

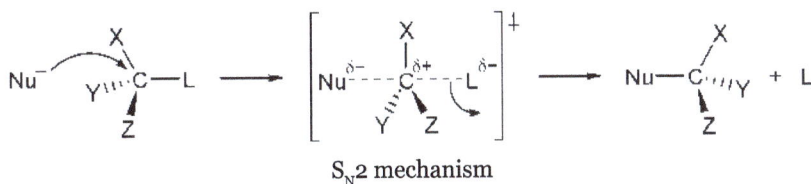

S_N2 mechanism

In the first type, a nucleophile, an atom or molecule with an excess of electrons and thus a negative charge or partial charge, replaces another atom or part of the "substrate" molecule. The electron pair from the nucleophile attacks the substrate forming a new bond, while the leaving group departs with an electron pair. The nucleophile may be electrically neutral or negatively charged, whereas the substrate is typically neutral or positively charged. Examples of nucleophiles are hydroxide ion, alkoxides, amines and halides. This type of reaction is found mainly in aliphatic hydrocarbons, and rarely in aromatic hydrocarbon. The latter have high electron density and enter nucleophilic aromatic substitution only with very strong electron withdrawing groups. Nucleophilic substitution can take place by two different mechanisms, S_N1 and S_N2. In their names, S stands for substitution, N for nucleophilic, and the number represents the kinetic order of the reaction, unimolecular or bimolecular.

The S_N1 reaction proceeds in two steps. First, the leaving group is eliminated creating a carbocation. This is followed by a rapid reaction with the nucleophile.

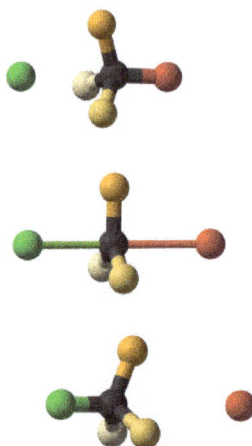

The three steps of an S_N2 reaction. The nucleophile is green and the leaving group is red

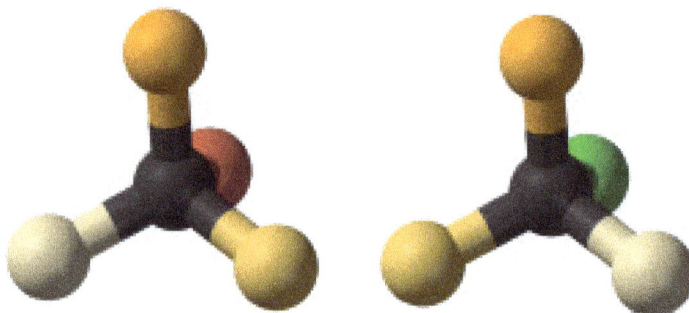

S_N2 reaction causes stereo inversion (Walden inversion)

In the S_N2 mechanism, the nucleophile forms a transition state with the attacked molecule, and only then the leaving group is cleaved. These two mechanisms differ in the stereochemistry of the products. S_N1 leads to the non-stereospecific addition and does not result in a chiral center, but rather in a set of geometric isomers (*cis/trans*). In contrast, a reversal (Walden inversion) of the previously existing stereochemistry is observed in the S_N2 mechanism.

Electrophilic substitution is the counterpart of the nucleophilic substitution in that the attacking atom or molecule, an electrophile, has low electron density and thus a positive charge. Typical electrophiles are the carbon atom of carbonyl groups, carbocations or sulfur or nitronium cations. This reaction takes place almost exclusively in aromatic hydrocarbons, where it is called electrophilic aromatic substitution. The electrophile attack results in the so-called σ-complex, a transition state in which the aromatic system is abolished. Then, the leaving group, usually a proton, is split off and the aromaticity is restored. An alternative to aromatic substitution is electrophilic aliphatic substitution. It is similar to the nucleophilic aliphatic substitution and also has two major types, S_E1 and S_E2

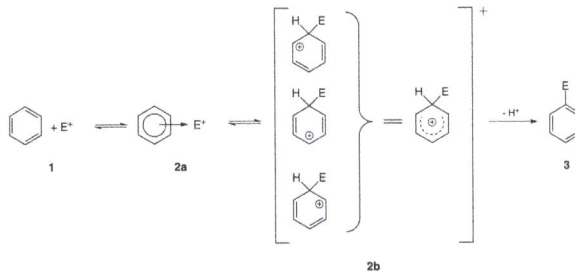

Mechanism of electrophilic aromatic substitution

In the third type of substitution reaction, radical substitution, the attacking particle is a radical. This process usually takes the form of a chain reaction, for example in the reaction of alkanes with halogens. In the first step, light or heat disintegrates the halogen-containing molecules producing the radicals. Then the reaction proceeds as an avalanche until two radicals meet and recombine.

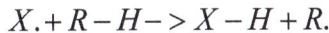

$$X. + R - H -> X - H + R.$$

$$R. + X2 -> R - X + X.$$

Reactions during the chain reaction of radical substitution

Addition and Elimination

The addition and its counterpart, the elimination, are reactions which change the number of substitutents on the carbon atom, and form or cleave multiple bonds. Double and triple bonds can be produced by eliminating a suitable leaving group. Similar to the nucleophilic substitution, there are several possible reaction mechanisms which are named after the respective reaction order. In the E1 mechanism, the leaving group is ejected first, forming a carbocation. The next step, formation of the double bond, takes place with elimination of a proton (deprotonation). The leaving order is reversed in the E1cb mechanism, that is the proton is split off first. This mechanism requires participation of a base. Because of the similar conditions, both reactions in the E1 or E1cb elimination always compete with the S_N1 substitution.

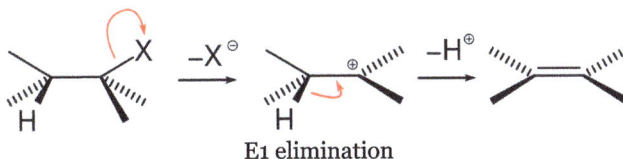

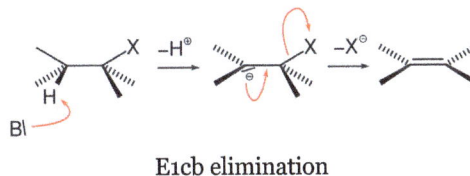

E1 elimination

E1cb elimination

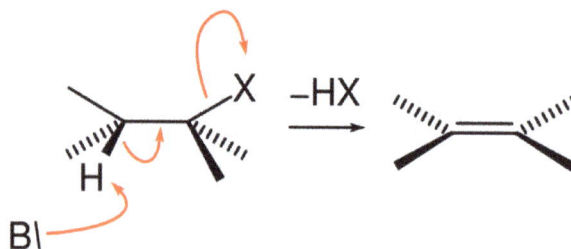

E2 elimination

The E2 mechanism also requires a base, but there the attack of the base and the elimination of the leaving group proceed simultaneously and produce no ionic intermediate. In contrast to the E1 eliminations, different stereochemical configurations are possible for the reaction product in the E2 mechanism, because the attack of the base preferentially occurs in the anti-position with respect to the leaving group. Because of the similar conditions and reagents, the E2 elimination is always in competition with the S_N2-substitution.

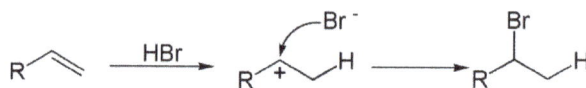

Electrophilic addition of hydrogen bromide

The counterpart of elimination is the addition where double or triple bonds are converted into single bonds. Similar to the substitution reactions, there are several types of additions distinguished by the type of the attacking particle. For example, in the electrophilic addition of hydrogen bromide, an electrophile (proton) attacks the double bond forming a carbocation, which then reacts with the nucleophile (bromine). The carbocation can be formed on either side of the double bond depending on the groups attached to its ends, and the preferred configuration can be predicted with the Markovnikov's rule. This rule states that "In the heterolytic addition of a polar molecule to an alkene or alkyne, the more electronegative (nucleophilic) atom (or part) of the polar molecule becomes attached to the carbon atom bearing the smaller number of hydrogen atoms."

If the addition of a functional group takes place at the less substituted carbon atom of the double bond, then the electrophilic substitution with acids is not possible. In this case, one has to use the hydroboration–oxidation reaction, where in the first step, the boron atom acts as electrophile and adds to the less substituted carbon atom. At the second step, the nucleophilic hydroperoxide or halogen anion attacks the boron atom.

While the addition to the electron-rich alkenes and alkynes is mainly electrophilic, the nucleophilic addition plays an important role for the carbon-heteroatom multiple bonds, and especially its most important representative, the carbonyl group. This process is often associated with an elimination, so that after the reaction the carbonyl group is present again. It is therefore called addition-elimination reaction and may occur in carboxylic acid derivatives such as chlorides, esters or anhydrides. This reaction is often catalyzed by acids or bases, where the acids increase by the electrophilicity of the carbonyl group by binding to the oxygen atom, whereas the bases enhance the nucleophilicity of the attacking nucleophile.

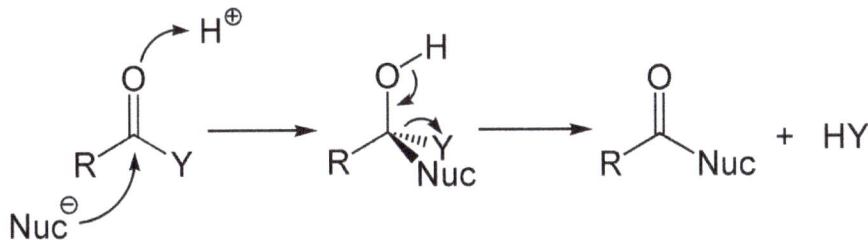

Acid-catalyzed addition-elimination mechanism

Nucleophilic addition of a carbanion or another nucleophile to the double bond of an alpha, beta unsaturated carbonyl compound can proceed via the Michael reaction, which belongs to the larger class of conjugate additions. This is one of the most useful methods for the mild formation of C–C bonds.

Some additions which can not be executed with nucleophiles and electrophiles, can be succeeded with free radicals. As with the free-radical substitution, the radical addition proceeds as a chain reaction, and such reactions are the basis of the free-radical polymerization.

Other Organic Reaction Mechanisms

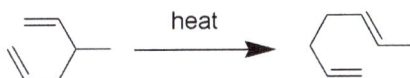

The Cope Rearrangement of 3-methyl-1,5-hexadiene

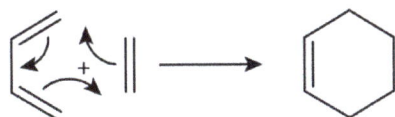

Mechanism of a Diels-Alder reaction

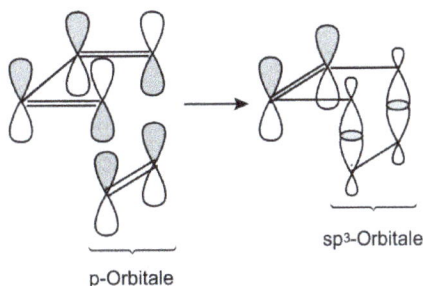

sp3-Orbitale

p-Orbitale

Orbital overlap in a Diels-Alder reaction

In a rearrangement reaction, the carbon skeleton of a molecule is rearranged to give a structural isomer of the original molecule. These include hydride shift reactions such as the Wagner-Meerwein rearrangement, where a hydrogen, alkyl or aryl group migrates from one carbon to a neighboring carbon. Most rearrangements are associated with the breaking and formation of new carbon-carbon bonds. Other examples are sigmatropic reaction such as the Cope rearrangement.

Cyclic rearrangements include cycloadditions and, more generally, pericyclic reactions, wherein two or more double bond-containing molecules form a cyclic molecule. An important example of cycloaddition reaction is the Diels–Alder reaction (the so-called [4+2] cycloaddition) between a conjugated diene and a substituted alkene to form a substituted cyclohexene system.

Whether a certain cycloaddition would proceed depends on the electronic orbitals of the participating species, as only orbitals with the same sign of wave function will overlap and interact constructively to form new bonds. Cycloaddition is usually assisted by light or heat. These perturbations result in different arrangement of electrons in the excited state of the involved molecules and therefore in different effects. For example, the [4+2] Diels-Alder reactions can be assisted by heat whereas the [2+2] cycloaddition is selectively induced by light. Because of the orbital character, the potential for developing stereoisomeric products upon cycloaddition is limited, as described by the Woodward–Hoffmann rules.

Biochemical Reactions

Biochemical reactions are mainly controlled by enzymes. These proteins can specifically catalyze a single reaction, so that reactions can be controlled very precisely. The reaction takes place in the active site, a small part of the enzyme which is usually found in a cleft or pocket lined by amino acid residues, and the rest of the enzyme is used mainly for stabilization. The catalytic action of enzymes relies on several mechanisms including the molecular shape ("induced fit"), bond strain, proximity and orientation of molecules relative to the enzyme, proton donation or withdrawal (acid/base catalysis), electrostatic interactions and many others.

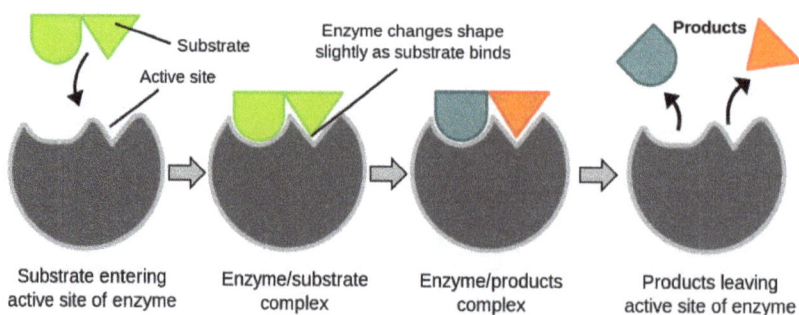

Illustration of the induced fit model of enzyme activity

The biochemical reactions that occur in living organisms are collectively known as metabolism. Among the most important of its mechanisms is the anabolism, in which different DNA and enzyme-controlled processes result in the production of large molecules such as proteins and carbohydrates from smaller units. Bioenergetics studies the sources of energy for such reactions. An important energy source is glucose, which can be produced by plants via photosynthesis or assimilated from food. All organisms use this energy to produce adenosine triphosphate (ATP), which can then be used to energize other reactions.

Applications

Chemical reactions are central to chemical engineering where they are used for the synthesis of new compounds from natural raw materials such as petroleum and mineral ores. It is essential to make the reaction as efficient as possible, maximizing the yield and minimizing the amount of reagents, energy inputs and waste. Catalysts are especially helpful for reducing the energy required for the reaction and increasing its reaction rate.

Thermite reaction proceeding in railway welding. Shortly after this, the liquid iron flows into the mould around the rail gap

Some specific reactions have their niche applications. For example, the thermite reaction is used to generate light and heat in pyrotechnics and welding. Although it is less controllable than the more conventional oxy-fuel welding, arc welding and flash welding, it requires much less equipment and is still used to mend rails, especially in remote areas.

Monitoring

Mechanisms of monitoring chemical reactions depend strongly on the reaction rate. Relatively slow processes can be analyzed in situ for the concentrations and identities of the individual ingredients. Important tools of real time analysis are the measurement of pH and analysis of optical absorption (color) and emission spectra. A less accessible but rather efficient method is introduction of a radioactive isotope into the reaction and monitoring how it changes over time and where it moves to; this method is often used to analyze redistribution of substances in the human body. Faster reactions are usually studied with ultrafast laser spectroscopy where utilization of femtosecond lasers allows short-lived transition states to be monitored at time scaled down to a few femtoseconds.

Catalysis

An air filter that utilizes low-temperature oxidation catalyst used to convert carbon monoxide to less toxic carbon dioxide at room temperature. It can also remove formaldehyde from the air.

Catalysis is the increase in the rate of a chemical reaction due to the participation of an additional substance called a catalyst. With a catalyst, reactions occur faster because they require less activation energy. Furthermore since they are not consumed in the catalyzed reaction, catalysts can continue to act repeatedly. Often only tiny amounts are required in principle.

Technical Perspective

In the presence of a catalyst, less free energy is required to reach the transition state, but the total free energy from reactants to products does not change. A catalyst may participate in multiple chemical transformations. The effect of a catalyst may vary due to the presence of other substances known as inhibitors or poisons (which reduce the catalytic activity) or promoters (which increase the activity). The opposite of a catalyst, a substance that reduces the rate of a reaction, is an inhibitor.

Catalyzed reactions have a lower activation energy (rate-limiting free energy of activation) than the corresponding uncatalyzed reaction, resulting in a higher reaction rate at the same temperature and for the same reactant concentrations. However, the detailed mechanics of catalysis is complex. Catalysts may affect the reaction environment favorably, or bind to the reagents to polarize bonds, e.g. acid catalysts for reactions of carbonyl compounds, or form specific intermediates that are not produced naturally, such as osmate esters in osmium tetroxide-catalyzed dihydroxylation of alkenes, or cause dissociation of reagents to reactive forms, such as chemisorbed hydrogen in catalytic hydrogenation.

Kinetically, catalytic reactions are typical chemical reactions; i.e. the reaction rate depends on the frequency of contact of the reactants in the rate-determining step. Usually, the catalyst participates in this slowest step, and rates are limited by amount of catalyst and its "activity". In heterogeneous catalysis, the diffusion of reagents to the surface and diffusion of products from the surface can be rate determining. A nanomaterial-based catalyst is an example of a heterogeneous catalyst. Analogous events associated with substrate binding and product dissociation apply to homogeneous catalysts.

Although catalysts are not consumed by the reaction itself, they may be inhibited, deactivated, or destroyed by secondary processes. In heterogeneous catalysis, typical secondary processes include coking where the catalyst becomes covered by polymeric side products. Additionally, heterogeneous catalysts can dissolve into the solution in a solid–liquid system or sublimate in a solid–gas system.

Background

The production of most industrially important chemicals involves catalysis. Similarly, most biochemically significant processes are catalysed. Research into catalysis is a major field in applied science and involves many areas of chemistry, notably organometallic chemistry and materials science. Catalysis is relevant to many aspects of environmental science, e.g. the catalytic converter in automobiles and the dynamics of the ozone hole. Catalytic reactions are preferred in environmentally friendly green chemistry due to the reduced amount of waste generated, as opposed to stoichiometric reactions in which all reactants are consumed and more side products are formed. Many transition metals and transition metal complexes are used in catalysis as well. Catalysts called enzymes are important in biology.

A catalyst works by providing an alternative reaction pathway to the reaction product. The rate of the reaction is increased as this alternative route has a lower activation energy than the reaction route not mediated by the catalyst. The disproportionation of hydrogen peroxide creates water and oxygen, as shown below.

$$2\,H_2O_2 \rightarrow 2\,H_2O + O_2$$

This reaction is preferable in the sense that the reaction products are more stable than the starting material, though the uncatalysed reaction is slow. In fact, the decomposition of hydrogen peroxide is so slow that hydrogen peroxide solutions are commercially available. This reaction is strongly affected by catalysts such as manganese dioxide, or the enzyme peroxidase in organisms. Upon the addition of a small amount of manganese dioxide, the hydrogen peroxide reacts rapidly. This effect is readily seen by the effervescence of oxygen. The manganese dioxide is not consumed in the reaction, and thus may be recovered unchanged, and re-used indefinitely. Accordingly, manganese dioxide *catalyses* this reaction.

General Principles

Units

Catalytic activity is usually denoted by the symbol z and measured in mol/s, a unit which was called katal and defined the SI unit for catalytic activity since 1999. Catalytic activity is not a kind of reaction rate, but a property of the catalyst under certain conditions, in relation to a specific chemical reaction. Catalytic activity of one katal (Symbol 1 kat = 1 mol/s) of a catalyst means an amount of that catalyst (substance, in Mol) that leads to a net reaction of one Mol per second of the reactants to the resulting reagents or other outcome which was intended for this chemical reaction. A catalyst may and usually will have different catalytic activity for distinct reactions.

Typical Mechanism

Catalysts generally react with one or more reactants to form intermediates that subsequently give the final reaction product, in the process regenerating the catalyst. The following is a typical reaction scheme, where C represents the catalyst, X and Y are reactants, and Z is the product of the reaction of X and Y:

$$X + C \rightarrow XC \ (1)$$

$$Y + XC \rightarrow XYC \ (2)$$

$$XYC \rightarrow CZ \ (3)$$

$$CZ \rightarrow C + Z \ (4)$$

Although the catalyst is consumed by reaction 1, it is subsequently produced by reaction 4, so it does not occur in the overall reaction equation:

$$X + Y \rightarrow Z$$

As a catalyst is regenerated in a reaction, often only small amounts are needed to increase the rate of the reaction. In practice, however, catalysts are sometimes consumed in secondary processes.

The catalyst does usually appear in the rate equation. For example, if the rate-determining step in the above reaction scheme is the first step

X + C → XC, the catalyzed reaction will be second order with rate equation $v = k_{cat}[X][C]$, which is proportional to the catalyst concentration [C]. However [C] remains constant during the reaction so that the catalyzed reaction is pseudo-first order: $v = k_{obs}[X]$, where $k_{obs} = k_{cat}[C]$.

As an example of a detailed mechanism at the microscopic level, in 2008 Danish researchers first revealed the sequence of events when oxygen and hydrogen combine on the surface of titanium dioxide (TiO_2, or *titania*) to produce water. With a time-lapse series of scanning tunneling microscopy images, they determined the molecules undergo adsorption, dissociation and diffusion before reacting. The intermediate reaction states were: HO_2, H_2O_2, then H_3O_2 and the final reaction product (water molecule dimers), after which the water molecule desorbs from the catalyst surface.

Reaction Energetics

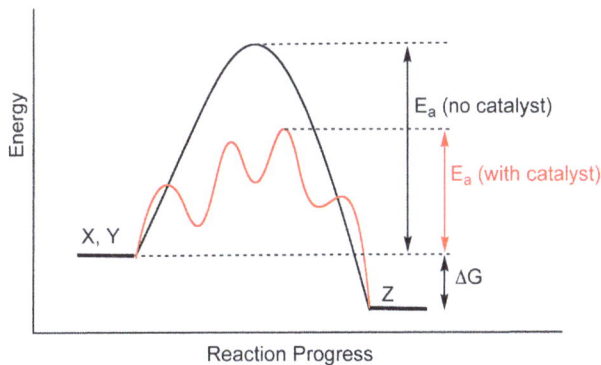

Generic potential energy diagram showing the effect of a catalyst in a hypothetical exothermic chemical reaction X + Y to give Z. The presence of the catalyst opens a different reaction pathway (shown in red) with a lower activation energy. The final result and the overall thermodynamics are the same.

Catalysts work by providing an (alternative) mechanism involving a different transition state and lower activation energy. Consequently, more molecular collisions have the energy needed to reach the transition state. Hence, catalysts can enable reactions that would otherwise be blocked or slowed by a kinetic barrier. The catalyst may increase reaction rate or selectivity, or enable the reaction at lower temperatures. This effect can be illustrated with an energy profile diagram.

In the catalyzed elementary reaction, catalysts do not change the extent of a reaction: they have no effect on the chemical equilibrium of a reaction because the rate of both the forward and the reverse reaction are both affected. The second law of thermodynamics describes why a catalyst does not change the chemical equilibrium of a reaction. Suppose there was such a catalyst that shifted an equilibrium. Introducing the catalyst to the system would result in a reaction to move to the new equilibrium, producing energy. Production of energy is a necessary result since reactions are spontaneous only if Gibbs free energy is produced, and if there is no energy barrier, there is no need for a catalyst. Then, removing the catalyst would also result in reaction, producing energy; i.e. the addition and its reverse process, removal, would both produce energy. Thus, a catalyst that could change the equilibrium would be a perpetual motion machine, a contradiction to the laws of thermodynamics.

If a catalyst does change the equilibrium, then it must be consumed as the reaction proceeds, and thus it is also a reactant. Illustrative is the base-catalysed hydrolysis of esters, where the produced carboxylic acid immediately reacts with the base catalyst and thus the reaction equilibrium is shifted towards hydrolysis.

The SI derived unit for measuring the catalytic activity of a catalyst is the katal, which is moles per second. The productivity of a catalyst can be described by the turn over number (or TON) and the catalytic activity by the *turn over frequency* (TOF), which is the TON per time unit. The biochemical equivalent is the enzyme unit. For more information on the efficiency of enzymatic catalysis.

The catalyst stabilizes the transition state more than it stabilizes the starting material. It decreases the kinetic barrier by decreasing the *difference* in energy between starting material and transition state. It does not change the energy difference between starting materials and products (thermodynamic barrier), or the available energy (this is provided by the environment as heat or light).

Materials

The chemical nature of catalysts is as diverse as catalysis itself, although some generalizations can be made. Proton acids are probably the most widely used catalysts, especially for the many reactions involving water, including hydrolysis and its reverse. Multifunctional solids often are catalytically active, e.g. zeolites, alumina, higher-order oxides, graphitic carbon, nanoparticles, nanodots, and facets of bulk materials. Transition metals are often used to catalyze redox reactions (oxidation, hydrogenation). Examples are nickel, such as Raney nickel for hydrogenation, and vanadium(V) oxide for oxidation of sulfur dioxide into sulfur trioxide by the so-called contact process. Many catalytic processes, especially those used in organic synthesis, require "late transition metals", such as palladium, platinum, gold, ruthenium, rhodium, or iridium.

Some so-called catalysts are really precatalysts. Precatalysts convert to catalysts in the reaction. For example, Wilkinson's catalyst $RhCl(PPh_3)_3$ loses one triphenylphosphine ligand before entering the true catalytic cycle. Precatalysts are easier to store but are easily activated in situ. Because of this preactivation step, many catalytic reactions involve an induction period.

Chemical species that improve catalytic activity are called co-catalysts (cocatalysts) or promotors in cooperative catalysis.

Types

Catalysts can be heterogeneous or homogeneous, depending on whether a catalyst exists in the same phase as the substrate. Biocatalysts (enzymes) are often seen as a separate group.

Heterogeneous Catalysts

Heterogeneous catalysts act in a different phase than the reactants. Most heterogeneous catalysts are solids that act on substrates in a liquid or gaseous reaction mixture. Diverse mechanisms for reactions on surfaces are known, depending on how the adsorption takes place (Langmuir-Hinshelwood, Eley-Rideal, and Mars-van Krevelen). The total surface area of solid has an important effect on the reaction rate. The smaller the catalyst particle size, the larger the surface area for a given mass of particles.

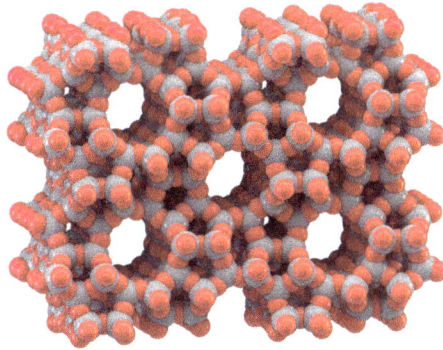

The microporous molecular structure of the zeolite ZSM-5 is exploited in catalysts used in refineries

Zeolites are extruded as pellets for easy handling in catalytic reactors.

A heterogeneous catalyst has active sites, which are the atoms or crystal faces where the reaction actually occurs. Depending on the mechanism, the active site may be either a planar exposed metal surface, a crystal edge with imperfect metal valence or a complicated combination of the two. Thus, not only most of the volume, but also most of the surface of a heterogeneous catalyst may be catalytically inactive. Finding out the nature of the active site requires technically challenging research. Thus, empirical research for finding out new metal combinations for catalysis continues.

For example, in the Haber process, finely divided iron serves as a catalyst for the synthesis of ammonia from nitrogen and hydrogen. The reacting gases adsorb onto active sites on the iron particles. Once physically adsorbed, the reagents undergo chemisorption that results in dissociation into adsorbed atomic species, and new bonds between the resulting fragments form in part due to their close proximity. In this way the particularly strong triple bond in nitrogen is broken, which would be extremely uncommon in the gas phase due to its high activation energy. Thus, the activation energy of the overall reaction is lowered, and the rate of reaction increases. Another place where a heterogeneous catalyst is applied is in the oxidation of sulfur dioxide on vanadium(V) oxide for the production of sulfuric acid.

Heterogeneous catalysts are typically "supported," which means that the catalyst is dispersed on a second material that enhances the effectiveness or minimizes their cost. Supports prevent or reduce agglomeration and sintering of the small catalyst particles, exposing more surface area, thus catalysts have a higher specific activity (per gram) on a support. Sometimes the support is merely a surface on which the catalyst is spread to increase the surface area. More often, the support and the catalyst interact, affecting the catalytic reaction. Supports are porous materials with a high surface area, most commonly alumina, zeolites or various kinds of activated carbon. Specialized supports include silicon dioxide, titanium dioxide, calcium carbonate, and barium sulfate.

Electrocatalysts

In the context of electrochemistry, specifically in fuel cell engineering, various metal-containing catalysts are used to enhance the rates of the half reactions that comprise the fuel cell. One common type of fuel cell electrocatalyst is based upon nanoparticles of platinum that are supported on slightly larger carbon particles. When in contact with one of the electrodes in a fuel cell, this platinum increases the rate of oxygen reduction either to water, or to hydroxide or hydrogen peroxide.

Homogeneous Catalysts

Homogeneous catalysts function in the same phase as the reactants, but the mechanistic principles invoked in heterogeneous catalysis are generally applicable. Typically homogeneous catalysts are dissolved in a solvent with the substrates. One example of homogeneous catalysis involves the influence of H^+ on the esterification of carboxylic acids, such as the formation of methyl acetate from acetic acid and methanol. For inorganic chemists, homogeneous catalysis is often synonymous with organometallic catalysts.

Organocatalysis

Whereas transition metals sometimes attract most of the attention in the study of catalysis, small organic molecules without metals can also exhibit catalytic properties, as is apparent from the fact that many enzymes lack transition metals. Typically, organic catalysts require a higher loading (amount of catalyst per unit amount of reactant, expressed in mol% amount of substance) than transition metal(-ion)-based catalysts, but these catalysts are usually commercially available in bulk, helping to reduce costs. In the early 2000s, these organocatalysts were considered "new generation" and are competitive to traditional metal(-ion)-containing catalysts. Organocatalysts are supposed to operate akin to metal-free enzymes utilizing, e.g., non-covalent interactions such as hydrogen bonding. The discipline organocatalysis is divided in the application of covalent (e.g., proline, DMAP) and non-covalent (e.g., thiourea organocatalysis) organocatalysts referring to the preferred catalyst-substrate binding and interaction, respectively.

Enzymes and Biocatalysts

In biology, enzymes are protein-based catalysts in metabolism and catabolism. Most biocatalysts are enzymes, but other non-protein-based classes of biomolecules also exhibit catalytic properties including ribozymes, and synthetic deoxyribozymes.

Biocatalysts can be thought of as intermediate between homogeneous and heterogeneous catalysts, although strictly speaking soluble enzymes are homogeneous catalysts and membrane-bound enzymes are heterogeneous. Several factors affect the activity of enzymes (and other catalysts) including temperature, pH, concentration of enzyme, substrate, and products. A particularly important reagent in enzymatic reactions is water, which is the product of many bond-forming reactions and a reactant in many bond-breaking processes.

In biocatalysis, enzymes are employed to prepare many commodity chemicals including high-fructose corn syrup and acrylamide.

Some monoclonal antibodies whose binding target is a stable molecule which resembles the transition state of a chemical reaction can function as weak catalysts for that chemical reaction by lowering its activation energy. Such catalytic antibodies are sometimes called "abzymes".

Nanocatalysts

Nanocatalysts are nanomaterials with catalytic activities. They have been extensively explored for wide range of applications. Among them, the nanocatalysts with enzyme mimicking activities are collectively called as nanozymes.

Tandem catalysis

In tandem catalysis two or more different catalysts are coupled in a one-pot reaction.

Autocatalysis

In autocatalysis the catalyst *is* a product of the overall reaction, in contrast to all other types of catalysis considered in this article. The simplest example of autocatalysis is a reaction of type A + B → 2 B, in one or in several steps. The overall reaction is just A → B, so that B is a product. But since B is also a reactant, it may be present in the rate equation and affect the reaction rate. As the reaction proceeds, the concentration of B increases and can accelerate the reaction as a catalyst. In effect, the reaction accelerates itself or is autocatalyzed.

A real example is the hydrolysis of an ester such as aspirin to a carboxylic acid and an alcohol. In the absence of added acid catalysts, the carboxylic acid product catalyzes the hydrolysis.

Significance

Estimates are that 90% of all commercially produced chemical products involve catalysts at some stage in the process of their manufacture. In 2005, catalytic processes generated about $900 billion in products worldwide. Catalysis is so pervasive that sub-areas are not readily classified. Some areas of particular concentration are surveyed below.

Left: Partially caramelised cube sugar, Right: burning cube sugar with ash as catalyst

Energy Processing

Petroleum refining makes intensive use of catalysis for alkylation, catalytic cracking (breaking long-chain hydrocarbons into smaller pieces), naphtha reforming and steam reforming (conversion of hydrocarbons into synthesis gas). Even the exhaust from the burning of fossil fuels is treated via catalysis: Catalytic converters, typically composed of platinum and rhodium, break down some of the more harmful byproducts of automobile exhaust.

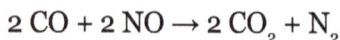

$$2\,CO + 2\,NO \rightarrow 2\,CO_2 + N_2$$

With regard to synthetic fuels, an old but still important process is the Fischer-Tropsch synthesis of hydrocarbons from synthesis gas, which itself is processed via water-gas shift reactions, catalysed by iron. Biodiesel and related biofuels require processing via both inorganic and biocatalysts.

Fuel cells rely on catalysts for both the anodic and cathodic reactions.

Catalytic heaters generate flameless heat from a supply of combustible fuel.

Bulk Chemicals

Some of the largest-scale chemicals are produced via catalytic oxidation, often using oxygen. Examples include nitric acid (from ammonia), sulfuric acid (from sulfur dioxide to sulfur trioxide by the chamber process), terephthalic acid from p-xylene, and acrylonitrile from propane and ammonia.

Many other chemical products are generated by large-scale reduction, often via hydrogenation. The largest-scale example is ammonia, which is prepared via the Haber process from nitrogen. Methanol is prepared from carbon monoxide.

Bulk polymers derived from ethylene and propylene are often prepared via Ziegler-Natta catalysis. Polyesters, polyamides, and isocyanates are derived via acid-base catalysis.

Most carbonylation processes require metal catalysts, examples include the Monsanto acetic acid process and hydroformylation.

Fine Chemicals

Many fine chemicals are prepared via catalysis; methods include those of heavy industry as well as more specialized processes that would be prohibitively expensive on a large scale. Examples include the Heck reaction, and Friedel-Crafts reactions.

Because most bioactive compounds are chiral, many pharmaceuticals are produced by enantioselective catalysis (catalytic asymmetric synthesis).

Food Processing

One of the most obvious applications of catalysis is the hydrogenation (reaction with hydrogen gas) of fats using nickel catalyst to produce margarine. Many other foodstuffs are prepared via biocatalysis.

Environment

Catalysis impacts the environment by increasing the efficiency of industrial processes, but catalysis also plays a direct role in the environment. A notable example is the catalytic role of chlorine free radicals in the breakdown of ozone. These radicals are formed by the action of ultraviolet radiation on chlorofluorocarbons (CFCs).

$$Cl^{\cdot} + O_3 \rightarrow ClO^{\cdot} + O_2$$

$$ClO^{\cdot} + O^{\cdot} \rightarrow Cl^{\cdot} + O_2$$

History

The concept of catalysis was invented by chemist Elizabeth Fulhame and described in a 1794 book, based on her novel work in oxidation-reduction experiments. The term *catalysis* was later used by Jöns Jakob Berzelius in 1835 to describe reactions that are accelerated by substances that remain unchanged after the reaction. Fulhame, who predated Berzelius, did work with water as opposed to metals in her reduction experiments. Other 18th century chemists who worked in catalysis were Eilhard Mitscherlich who referred to it as *contact* processes, and Johann Wolfgang Döbereiner who spoke of *contact action*. He developed Döbereiner's lamp, a lighter based on hydrogen and a platinum sponge, which became a commercial success in the 1820s that lives on today. Humphry Davy discovered the use of platinum in catalysis. In the 1880s, Wilhelm Ostwald at Leipzig University started a systematic investigation into reactions that were catalyzed by the presence of acids and bases, and found that chemical reactions occur at finite rates and that these rates can be used to determine the strengths of acids and bases. For this work, Ostwald was awarded the 1909 Nobel Prize in Chemistry.

Inhibitors, Poisons, and Promoters

Substances that reduce the action of catalysts are called catalyst inhibitors if reversible, and catalyst poisons if irreversible. Promoters are substances that increase the catalytic activity, even though they are not catalysts by themselves.

Inhibitors are sometimes referred to as "negative catalysts" since they decrease the reaction rate. However the term inhibitor is preferred since they do not work by introducing a reaction path with higher activation energy; this would not reduce the rate since the reaction would continue to occur by the non-catalyzed path. Instead they act either by deactivating catalysts, or by removing reaction intermediates such as free radicals.

The inhibitor may modify selectivity in addition to rate. For instance, in the reduction of alkynes to alkenes, a palladium (Pd) catalyst partly "poisoned" with lead(II) acetate $(Pb(CH_3CO_2)_2)$ can be used. Without the deactivation of the catalyst, the alkene produced would be further reduced to alkane.

The inhibitor can produce this effect by, e.g., selectively poisoning only certain types of active sites. Another mechanism is the modification of surface geometry. For instance, in hydrogenation operations, large planes of metal surface function as sites of hydrogenolysis catalysis while sites catalyzing hydrogenation of unsaturates are smaller. Thus, a poison that covers surface randomly will tend to reduce the number of uncontaminated large planes but leave proportionally more smaller sites free, thus changing the hydrogenation vs. hydrogenolysis selectivity. Many other mechanisms are also possible.

Promoters can cover up surface to prevent production of a mat of coke, or even actively remove such material (e.g., rhenium on platinum in platforming). They can aid the dispersion of the catalytic material or bind to reagents.

Current Market

The global demand on catalysts in 2010 was estimated at approximately 29.5 billion USD. With the rapid recovery in automotive and chemical industry overall, the global catalyst market is expected to experience fast growth in the next years.

Industrial Catalysts

The first time a catalyst was used in the industry was in 1746 by J. Roebuck in the manufacture of lead chamber sulfuric acid. Since then catalysts have been in use in a large portion of the chemical industry. In the start only pure components were used as catalysts, but after the year 1900 multicomponent catalysts were studied and are now commonly used in the industry.

In the chemical industry and industrial research, catalysis play an important role. Different catalysts are in constant development to fulfill economic, political and environmental demands. When using a catalyst it is possible to replace a polluting chemical reaction with a more environmentally friendly alternative. Today, and in the future, this can be vital for the chemical industry. In addition it's important for a company/researcher to pay attention to market development. If a company's catalyst is not continually improved, another company can make progress in research on that particular catalyst and gain market share. For a company, a new and improved catalyst can be a huge advantage for a competitive manufacturing cost. It's extremely expensive for a company to shut down the plant because of an error in the catalyst, so the correct selection of a catalyst or a new improvement can be key to industrial success.

To achieve the best understanding and development of a catalyst it is important that different special fields work together. These fields can be: organic chemistry, analytic chemistry, inorganic chemistry, chemical engineers and surface chemistry. The economics must also be taken into account. One of the issues that must be considered is if the company should use money on doing the catalyst research themselves or buy the technology from someone else. As the analytical tools are becoming more advanced, the catalysts used in the industry are improving. One example of an improvement can be to develop a catalyst with a longer lifetime than the previous version. Some of the advantages an improved catalyst gives, that affects people's lives, are: cheaper and more effective fuel, new drugs and medications and new polymers.

Some of the large chemical processes that use catalysis today are the production of methanol and ammonia. Both methanol and ammonia synthesis take advantage of the water-gas shift reaction and heterogeneous catalysis, while other chemical industries use homogenous catalysis. If the catalyst exists in the same phase as the reactants it is said to be homogenous; otherwise it is heterogeneous.

Water Gas Shift Reaction

The water gas shift reaction was first used industrially at the beginning of the 20th century. Today the WGS reaction is used primarily to produce hydrogen that can be used for further production of methanol and ammonia.

WGS reaction:

(1) $CO + H_2O \leftrightarrow H_2 + CO_2$

The reaction refers to carbon monoxide (CO) that reacts with water (H_2O) to form carbon dioxide (CO_2) and hydrogen (H_2). The reaction is exothermic with ΔH= -41.1 kJ/mol and have an adiabatic temperature rise of 8–10 °C per percent CO converted to CO_2 and H_2.

The most common catalysts used in the water-gas shift reaction are the high temperature shift (HTS) catalyst and the low temperature shift (LTS) catalyst. The HTS catalyst consists of iron oxide stabilized by chromium oxide, while the LTS catalyst is based on copper. The main purpose of the LTS catalyst is to reduce CO content in the reformate which is especially important in the ammonia production for high yield of H_2. Both catalysts are necessary for thermal stability, since using the LTS reactor alone increases exit-stream temperatures to unacceptable levels.

The equilibrium constant for the reaction is given as:

(2) $K_p = (p_{H2} \times p_{CO2}) / (p_{CO} \times p_{H2O})$

(3) $K_p = e^{((4577.8K/T-4.22))}$

Low temperatures will therefore shift the reaction to the right, and more products will be produced. The equilibrium constant is extremely dependent on the reaction temperature, for example is the Kp equal to 228 at 200 °C, but only 11.8 at 400 °C. The WGS reaction can be performed both homogenously and heterogeneously, but only the heterogeneously way is used commercially.

High Temperature Shift (HTS) Catalyst

The first step in the WGS reaction is the high temperature shift which is carried out at temperatures between 320 °C and 450 °C. As mentioned before, the catalyst is a composition of iron-oxide, Fe_2O_3 (90-95%), and chromium oxides Cr_2O_3 (5-10%) which have an ideal activity and selectivity at these temperatures. When preparing this catalyst, one of the most important step is washing to remove sulfate that can turn into hydrogen sulfide and poison the LTS catalyst later in the process. Chromium is added to the catalyst to stabilize the catalyst activity over time and to delay sintering of iron oxide. Sintering will decrease the active catalyst area, so by decreasing the sintering rate the lifetime of the catalyst will be extended. The catalyst is usually used in pellets form, and the size play an important role. Large pellets will be strong, but the reaction rate will be limited.

In the end, the dominate phase in the catalyst consist of Cr_3+ in α-Fe_2O_3 but the catalyst is still not active. To be active α-Fe_2O_3 must be reduced to Fe and CrO_3 must be reduced to Cr in presence of H_2. This usually happens in the reactor start-up phase and because the reduction reactions are exothermic the reduction should happen under controlled circumstances. The lifetime of the iron-chrome catalyst is approximately 3–5 years, depending on how the catalyst is handled.

Even though the mechanism for the HTS catalyst has been done a lot of research on, there is no final agreement on the kinetics/mechanism. Research has narrowed it down to two possible mechanisms: a regenerative redox mechanism and an adsorptive(associative) mechanism.

The redox mechanism is given below:

First a CO molecule reduces an O molecule, yielding CO_2 and a vacant surface center:

$$(4)\ CO+(O) \rightarrow CO_2 + (*)$$

The vacant side is then reoxidized by water, and the oxide center is regenerated:

$$(5)\ H_2O+(*) \rightarrow H_2+ (O)$$

The adsorptive mechanism assumes that format species is produced when an adsorbed CO molecule reacts with a surface hydroxyl group:

$$(6)\ H_2O \rightarrow OH(ads)+ H(ads)$$

$$(7)\ CO(ads)+ OH(ads) \rightarrow COOH\ (ads)$$

The format decomposes then in the presence of steam:

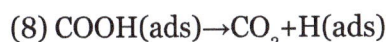

$$(8)\ COOH(ads) \rightarrow CO_2+H(ads)$$

$$(9)\ 2H(ads) \rightarrow H_2$$

Low Temperature Shift (LTS) Catalyst

The low temperature process is the second stage in the process, and is designed to take advantage of higher hydrogen equilibrium at low temperatures. The reaction is carried out between 200 °C and 250 °C, and the most commonly used catalyst is based on copper. While the HTS reactor used an iron-chrome based catalyst, the copper-catalyst is more active at lower temperatures thereby yielding a lower equilibrium concentration of CO and a higher equilibrium concentration of H_2. The disadvantage with a copper catalysts is that it is very sensitive when it comes to sulfide poisoning, a future use of for example a cobalt- molybdenum catalyst could solve this problem. The catalyst mainly used in the industry today is a copper-zinc-alumina ($Cu/ZnO/Al_2O_3$) based catalyst.

Also the LTS catalyst has to be activated by reduction before it can be used. The reduction reaction $CuO + H_2 \rightarrow Cu + H_2O$ is highly exothermic and should be conducted in dry gas for an optimal result.

As for the HTS catalyst mechanism, two similar reaction mechanisms are suggested. The first mechanism that was proposed for the LTS reaction was a redox mechanism, but later evidence showed that the reaction can proceed via associated intermediates. The different intermediates that is suggested are: HOCO, HCO and HCOO. In 2009 there are in total three mechanisms that are proposed for the water-gas shift reaction over Cu(111), given below.

Intermediate mechanism (usually called associative mechanism): An intermediate is first formed and then decomposes into the final products:

(10) CO + (species derived from H_2O) $\rightarrow$Intermediate$\rightarrow CO_2$

Associative mechanism: CO_2 produced from the reaction of CO with OH without the formation of an intermediate:

(11) CO + OH $\rightarrow$H + CO_2

Redox mechanism: Water dissociation that yields surface oxygen atoms which react with CO to produce CO_2:

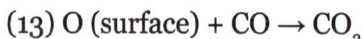

(12) $H_2O \rightarrow O$ (surface)

(13) O (surface) + CO $\rightarrow CO_2$

It is not said that just one of these mechanisms is controlling the reaction, it is possible that several of them are active. Q.-L. Tang *et al.* has suggested that the most favorable mechanism is the intermediate mechanism (with HOCO as intermediate) followed by the redox mechanism with the rate determining step being the water dissociation.

For both HTS catalyst and LTS catalyst the redox mechanism is the oldest theory and most published articles support this theory, but as technology has developed the adsorptive mechanism has become more of interest. One of the reasons to the fact that the literature is not agreeing on one mechanism can be because of experiments are carried out under different assumptions.

Carbon Monoxide

CO must be produced for the WGS reaction to take place. This can be done in different ways from a variety of carbon sources such as:

-passing steam over coal:

$$(14)\ C + H_2O = CO + H_2$$

-steam reforming methane, over a nickel catalyst:

$$(15)\ CH_4 + H_2O = CO + 3H_2$$

-or by using biomass. Both the reactions shown above are highly endothermic and can be coupled to an exothermic partial oxidation. The products of CO and H_2 are known as syngas.

When dealing with a catalyst and CO, it is common to assume that the intermediate CO-Metal is formated before the intermediate reacts further into the products. When designing a catalyst this is important to remember. The strength of interaction between the CO molecule and the metal should be strong enough to provide a sufficient concentration of the intermediate, but not so strong that the reaction will not continue.

CO is a common molecule to use in a catalytic reaction, and when it interacts with a metal surface it is actually the molecular orbitals of CO that interacts with the d-band of the metal surface. When considering a molecular orbital(MO)-diagram CO can act as an σ-donor via the lone pair of the electrons on C, and a π-acceptor ligand in transition metal complexes. When a CO molecule is adsorbed on a metal surface, the d-band of the metal will interact with the molecular orbitals of CO. It is possible to look at a simplified picture, and only consider the LUMO ($2\pi^*$) and HOMO (5σ) to CO. The overall effect of the σ-donation and the π- back donation is that a strong bond between C and the metal is being formed and in addition the bond between C and O will be weakened. The latter effect is due to charge depletion of the CO 5σ bonding and charge increase of the CO $2\pi^*$ antibonding orbital.

When looking at chemical surfaces, many researchers seems to agree on that the surface of the $Cu/Al_2O_3/ZnO$ is most similar to the Cu(111) surface. Since copper is the main catalyst and the active phase in the LTS catalyst, many experiments has been done with copper. In the figure given here experiments has been done on Cu(110) and Cu(111). The figure shows Arrhenius plot derived from reaction rates. It can be seen from the figure that Cu(110) shows a faster reaction rate and a lower activation energy. This can be due to the fact that Cu(111) is more closely packed than Cu(110).

Methanol Production

Production of methanol is an important industry today and methanol is one of the largest volume carbonylation products. The process uses syngas as feedstock and for that reason the water gas shift reaction is important for this synthesis. The most important reaction based on methanol is the decomposition of methanol to yield carbon monoxide and hydrogen. Methanol is therefore an important raw material for production of CO and H_2 that can be used in generation of fuel.

BASF was the first company (in 1923) to produce methanol on large-scale, then using a sulfur-resistant ZnO/Cr_2O_3 catalyst. The feed gas was produced by gasification over coal. Today the synthesis gas is usually manufactured via steam reforming of natural gas. The most effective catalysts for methanol synthesis are Cu, Ni, Pd and Pt, while the most common metals used for support are Al and Si. In 1966 ICI (Imperial Chemical Industries) developed a process that is still in use today. The process is a low-pressure process that uses a $Cu/ZnO/Al_2O_3$ catalyst where copper is the active material. This catalyst is actually the same that the low-temperature shift catalyst in the WGS reaction is using. The reaction described below is carried out at 250 °C and 5-10 MPa:

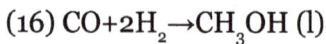

(16) $CO + 2H_2 \rightarrow CH_3OH$ (l)

(17) $CO_2 + 3H_2 \rightarrow CH_3OH$ (l) $+ H_2O$ (l)

Both of these reactions are exothermic and proceeds with volume contraction. Maximum yield of methanol is therefore obtained at low temperatures and high pressure and with use of a catalyst that has a high activity at these conditions. A catalyst with sufficiently high activity at the low temperature does still not exist, and this is one of the main reasons that companies keep doing research and catalyst development.

A reaction mechanism for methanol synthesis has been suggested by Chinchen *et al.*:

(18) $CO_2 \rightarrow CO_2^*$

(19) $H_2 \rightarrow 2H^*$

(20) $CO_2^* + H^* \rightarrow HCOO^*$

(21) $HCOO^* + 3H^* \rightarrow CH_3OH + O^*$

(22) $CO + O^* \rightarrow CO_2$

(23) $H_2 + O^* \rightarrow H_2O$

Today there are four different ways to catalytically obtain hydrogen production from methanol, and all reactions can be carried out by using a transition metal catalyst (Cu, Pd):

Steam Reforming

The reaction is given as:

(24) $CH_3OH(l) + H_2O$ (l) $\rightarrow CO_2 + 3H_2$ $\Delta H = +131$ KJ/mol

Steam reforming is a good source for production of hydrogen, but the reaction is endothermic. The reaction can be carried out over a copper-based catalyst, but the reaction mechanism is dependent on the catalyst. For a copper-based catalyst two different reaction mechanisms have been proposed, a decomposition-water-gas shift sequence and a mechanism that proceeds via methanol dehydrogenation to methyl formate. The first mechanism aims at methanol decomposition followed by the WGS reaction and has been proposed for the $Cu/ZnO/Al_2O_3$:

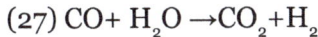

(25) $CH_3OH + H_2O \rightarrow CO_2 + 3H_2$

(26) $CH_3OH \rightarrow CO + 2H_2$

(27) $CO + H_2O \rightarrow CO_2 + H_2$

The mechanism for the methyl format reaction can be dependent of the composition of the catalyst. The following mechanism has been proposed over $Cu/ZnO/Al_2O_3$:

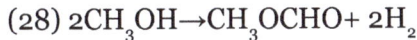

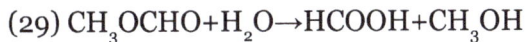

(28) $2CH_3OH \rightarrow CH_3OCHO + 2H_2$

(29) $CH_3OCHO + H_2O \rightarrow HCOOH + CH_3OH$

(30) $HCOOC \rightarrow CO_2 + H_2$

When methanol is almost completely converted CO is being produced as a secondary product via the reverse water-gas shift reaction.

Methanol Decomposition

The second way to produce hydrogen from methanol is by methanol decomposition:

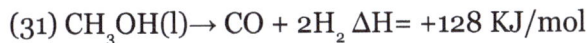

(31) $CH_3OH(l) \rightarrow CO + 2H_2 \ \Delta H = +128 \ KJ/mol$

As the enthalpy shows, the reaction is endothermic and this can be taken further advantage of in the industry. This reaction is the opposite of the methanol synthesis from syngas, and the most effective catalysts seems to be Cu, Ni, Pd and Pt as mentioned before. Often, a Cu/ZnO-based catalyst is used at temperatures between 200 and 300 °C but a production of by-product as dimethyl ether, methyl format, methane and water is common. The reaction mechanism is not fully understood and there are two possible mechanism proposed (2002) : one producing CO_2 and H_2 by decomposition of formate intermediates and the other producing CO and H_2 via a methyl formate intermediate.

Partial Oxidation

Partial oxidation is a third way for producing hydrogen from methanol. The reaction is given below, and is often carried out with air or oxygen as oxidant :

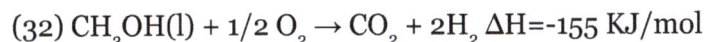

(32) $CH_3OH(l) + 1/2 \ O_2 \rightarrow CO_2 + 2H_2 \ \Delta H = -155 \ KJ/mol$

The reaction is exothermic and has, under favorable conditions, a higher reaction rate than steam reforming. The catalyst used is often Cu (Cu/ZnO) or Pd and they differ in qualities such as by-product formation, product distribution and the effect of oxygen partial pressure.

Combined Reforming

Combined reforming is a combination of partial oxidation and steam reforming and is the last reaction that is used for hydrogen production. The general equation is given below:

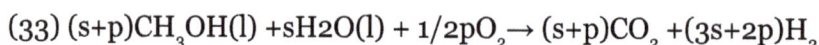

$$(33)\ (s+p)CH_3OH(l) + sH2O(l) + 1/2pO_2 \rightarrow (s+p)CO_2 + (3s+2p)H_2$$

s and p are the stoichiometric coefficients for steam reforming and partial oxidation, respectively. The reaction can be both endothermic and exothermic determined by the conditions, and combine both the advantages of steam reforming and partial oxidation.

Ammonia Synthesis

Ammonia synthesis was discovered by Fritz Haber, by using iron catalysts. The ammonia synthesis advanced between 1909 and 1913, and two important concepts were developed; the benefits of a promoter and the poisoning effect.

Ammonia production was one of the first commercial processes that required the production of hydrogen, and the cheapest and best way to obtain hydrogen was via the water-gas shift reaction. The Haber–Bosch process is the most common process used in the ammonia industry.

A lot of research has been done on the catalyst used in the ammonia process, but the main catalyst that is used today is not that dissimilar to the one that was first developed. The catalyst the industry use is a promoted iron catalyst, where the promoters can be K_2O (potassium oxide), Al_2O_3 (aluminium oxide) and CaO (calcium oxide) and the basic catalytic material is Fe. The most common is to use fixed bed reactors for the synthesis catalyst.

The main ammonia reaction is given below:

$$(34)\ N_2 + 3H_2 \leftrightarrow 2NH_3$$

The produced ammonia can be used further in production of nitric acid via the Ostwald process.

Polymerization

IUPAC Definition

polymerization: The process of converting a monomer or a mixture of monomers into a polymer.

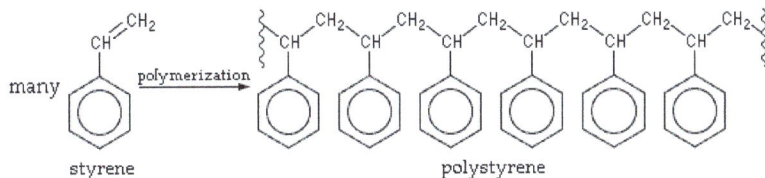

styrene   polystyrene

An example of alkene polymerization, in which each styrene monomer's double bond reforms as a single bond plus a bond to another styrene monomer. The product is polystyrene.

In polymer chemistry, polymerization is a process of reacting monomer molecules together in a chemical reaction to form polymer chains or three-dimensional networks. There are many forms of polymerization and different systems exist to categorize them.

Introduction

In chemical compounds, polymerization occurs via a variety of reaction mechanisms that vary in complexity due to functional groups present in reacting compounds and their inherent steric effects. In more straightforward polymerization, alkenes, which are relatively stable due to σ bonding between carbon atoms, form polymers through relatively simple radical reactions; in contrast, more complex reactions such as those that involve substitution at the carbonyl group require more complex synthesis due to the way in which reacting molecules polymerize. Alkanes can also be polymerized, but only with the help of strong acids.

As alkenes can be formed in somewhat straightforward reaction mechanisms, they form useful compounds such as polyethylene and polyvinyl chloride (PVC) when undergoing radical reactions, which are produced in high tonnages each year due to their usefulness in manufacturing processes of commercial products, such as piping, insulation and packaging. In general, polymers such as PVC are referred to as "homopolymers," as they consist of repeated long chains or structures of the same monomer unit, whereas polymers that consist of more than one molecule are referred to as copolymers (or co-polymers).

Other monomer units, such as formaldehyde hydrates or simple aldehydes, are able to polymerize themselves at quite low temperatures (ca. −80 °C) to form trimers; molecules consisting of 3 monomer units, which can cyclize to form ring cyclic structures, or undergo further reactions to form tetramers, or 4 monomer-unit compounds. Further compounds either being referred to as oligomers in smaller molecules. Generally, because formaldehyde is an exceptionally reactive electrophile it allows nucleophillic addition of hemiacetal intermediates, which are in general short-lived and relatively unstable "mid-stage" compounds that react with other molecules present to form more stable polymeric compounds.

Polymerization that is not sufficiently moderated and proceeds at a fast rate can be very hazardous. This phenomenon is known as hazardous polymerization and can cause fires and explosions.

Step-growth

Step-growth polymers are defined as polymers formed by the stepwise reaction between functional groups of monomers, usually containing heteroatoms such as nitrogen or oxygen. Most step-growth polymers are also classified as condensation polymers, but not all step-growth polymers (like polyurethanes formed from isocyanate and alcohol bifunctional monomers) release condensates; in this case, we talk about addition polymers. Step-growth polymers increase in molecular weight at a very slow rate at lower conversions and reach moderately high molecular weights only at very high conversion (i.e., >95%).

To alleviate inconsistencies in these naming methods, adjusted definition for condensation and addition polymers have been developed. A condensation polymer is defined as a polymer that involves loss of small molecules during its synthesis, or contains heteroatoms as part of its backbone chain, or its repeat unit does not contain all the atoms present in the hypothetical monomer to which it can be degraded.

Chain-growth

Chain-growth polymerization (or addition polymerization) involves the linking together of molecules incorporating double or triple carbon-carbon bonds. These unsaturated *monomers* (the identical molecules that make up the polymers) have extra internal bonds that are able to break and link up with other monomers to form a repeating chain, whose backbone typically contains only carbon atoms. Chain-growth polymerization is involved in the manufacture of polymers such as polyethylene, polypropylene, and polyvinyl chloride (PVC). A special case of chain-growth polymerization leads to living polymerization.

In the radical polymerization of ethylene, its π bond is broken, and the two electrons rearrange to create a new propagating center like the one that attacked it. The form this propagating center takes depends on the specific type of addition mechanism. There are several mechanisms through which this can be initiated. The free radical mechanism is one of the first methods to be used. Free radicals are very reactive atoms or molecules that have unpaired electrons. Taking the polymerization of ethylene as an example, the free radical mechanism can be divided into three stages: chain initiation, chain propagation, and chain termination.

Polymerization of ethylene

Free radical addition polymerization of ethylene must take place at high temperatures and pressures, approximately 300 °C and 2000 atm. While most other free radical polymerizations do not require such extreme temperatures and pressures, they do tend to lack control. One effect of this lack of control is a high degree of branching. Also, as termination occurs randomly, when two chains collide, it is impossible to control the length of individual chains. A newer method of polymerization similar to free radical, but allowing more control involves the Ziegler-Natta catalyst, especially with respect to polymer branching.

Other forms of chain growth polymerization include cationic addition polymerization and anionic addition polymerization. While not used to a large extent in industry yet due to stringent reaction conditions such as lack of water and oxygen, these methods provide ways to polymerize some monomers that cannot be polymerized by free radical methods such as polypropylene. Cationic and anionic mechanisms are also more ideally suited for living polymerizations, although free radical living polymerizations have also been developed.

Esters of acrylic acid contain a carbon-carbon double bond which is conjugated to an ester group. This allows the possibility of both types of polymerization mechanism. An acrylic ester by itself can undergo chain-growth polymerization to form a homopolymer with a carbon-carbon backbone, such as poly(methyl methacrylate). Also, however, certain acrylic esters can react with diamine monomers by nucleophilic conjugate addition of amine groups to acrylic C=C bonds. In this case the polymerization proceeds by step-growth and the products are poly(beta-amino ester) copolymers, with backbones containing nitrogen (as amine) and oxygen (as ester) as well as carbon.

Physical Polymer Reaction Engineering

To produce a high-molecular-weight, uniform product, various methods are employed to better control the initiation, propagation, and termination rates during chain polymerization and also to remove excess concentrated heat during these exothermic reactions compared to polymerization of the pure monomer (also referred to as bulk polymerization). These include emulsion polymerization, solution polymerization, suspension polymerization, and precipitation polymerization. Although the polymer polydispersity and molecular weight may be improved, these methods may introduce additional processing requirements to isolate the product from a solvent.

Photopolymerization

Most photopolymerization reactions are chain-growth polymerizations which are initiated by the absorption of visible or ultraviolet light. The light may be absorbed either directly by the reactant monomer (*direct* photopolymerization), or else by a *photosensitizer* which absorbs the light and then transfers energy to the monomer. In general only the initiation step differs from that of the ordinary thermal polymerization of the same monomer; subsequent propagation, termination and chain transfer steps are unchanged. In step-growth photopolymerization, absorption of light triggers an addition (or condensation) reaction between two comonomers that do not react without light. A propagation cycle is not initiated because each growth step requires the assistance of light.

Photopolymerization can be used as a photographic or printing process, because polymerization only occurs in regions which have been exposed to light. Unreacted monomer can be removed from unexposed regions, leaving a relief polymeric image. Several forms of 3D printing—including layer-by-layer stereolithography and two-photon absorption 3D photopolymerization—use photopolymerization.

Chemical Equilibrium

In a chemical reaction, chemical equilibrium is the state in which both reactants and products are present in concentrations which have no further tendency to change with time. Usually, this state results when the forward reaction proceeds at the same rate as the reverse reaction. The reaction rates of the forward and backward reactions are generally not zero, but equal. Thus, there are no net changes in the concentrations of the reactant(s) and product(s). Such a state is known as dynamic equilibrium.

Historical Introduction

Burette, a common laboratory apparatus for carrying out titration, an important experimental technique in equilibrium and analytical chemistry.

The concept of chemical equilibrium was developed after Berthollet (1803) found that some chemical reactions are reversible. For any reaction mixture to exist at equilibrium, the rates of the forward and backward (reverse) reactions are equal. In the following chemical equation with arrows pointing both ways to indicate equilibrium, A and B are reactant chemical species, S and T are product species, and α, β, σ, and τ are the stoichiometric coefficients of the respective reactants and products:

$$\alpha\,A + \beta\,B \rightleftharpoons \sigma\,S + \tau\,T$$

The equilibrium concentration position of a reaction is said to lie "far to the right" if, at equilibrium, nearly all the reactants are consumed. Conversely the equilibrium position is said to be "far to the left" if hardly any product is formed from the reactants.

Guldberg and Waage (1865), building on Berthollet's ideas, proposed the law of mass action:

$$\text{forward reaction rate} = k_+ A^\alpha B^\beta$$

$$\text{backward reaction rate} = k_- S^\sigma T^\tau$$

where A, B, S and T are active masses and k_+ and k_- are rate constants. Since at equilibrium forward and backward rates are equal:

$$k_+ \{A\}^\alpha \{B\}^\beta = k_- \{S\}^\sigma \{T\}^\tau$$

and the ratio of the rate constants is also a constant, now known as an equilibrium constant.

$$K_c = \frac{k_+}{k_-} = \frac{\{S\}^\sigma \{T\}^\tau}{\{A\}^\alpha \{B\}^\beta}$$

By convention the products form the numerator. However, the law of mass action is valid only for concerted one-step reactions that proceed through a single transition state and is not valid in general because rate equations do not, in general, follow the stoichiometry of the reaction as Guldberg and Waage had proposed (for example, nucleophilic aliphatic substitution by S_N1 or reaction of hydrogen and bromine to form hydrogen bromide). Equality of forward and backward reaction rates, however, is a necessary condition for chemical equilibrium, though it is not sufficient to explain why equilibrium occurs.

Despite the failure of this derivation, the equilibrium constant for a reaction is indeed a constant, independent of the activities of the various species involved, though it does depend on temperature as observed by the van 't Hoff equation. Adding a catalyst will affect both the forward reaction and the reverse reaction in the same way and will not have an effect on the equilibrium constant. The catalyst will speed up both reactions thereby increasing the speed at which equilibrium is reached.

Although the macroscopic equilibrium concentrations are constant in time, reactions do occur at the molecular level. For example, in the case of acetic acid dissolved in water and forming acetate and hydronium ions,

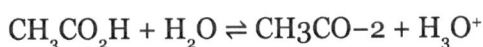

$$CH_3CO_2H + H_2O \rightleftharpoons CH3CO-2 + H_3O^+$$

a proton may hop from one molecule of acetic acid on to a water molecule and then on to an acetate anion to form another molecule of acetic acid and leaving the number of acetic acid molecules unchanged. This is an example of dynamic equilibrium. Equilibria, like the rest of thermodynamics, are statistical phenomena, averages of microscopic behavior.

Le Châtelier's principle (1884) gives an idea of the behavior of an equilibrium system when changes to its reaction conditions occur. *If a dynamic equilibrium is disturbed by changing the conditions, the position of equilibrium moves to partially reverse the change.* For example, adding more S from the outside will cause an excess of products, and the system will try to counteract this by increasing the reverse reaction and pushing the equilibrium point backward (though the equilibrium constant will stay the same).

If mineral acid is added to the acetic acid mixture, increasing the concentration of hydronium ion, the amount of dissociation must decrease as the reaction is driven to the left in accordance with this principle. This can also be deduced from the equilibrium constant expression for the reaction:

$$K = \frac{\{CH3CO2^-\}\{H3O+\}}{\{CH3CO2H\}}$$

If $\{H_3O^+\}$ increases $\{CH_3CO_2H\}$ must increase and CH3CO-2 must decrease. The H_2O is left out, as it is the solvent and its concentration remains high and nearly constant.

A quantitative version is given by the reaction quotient.

J. W. Gibbs suggested in 1873 that equilibrium is attained when the Gibbs free energy of the system is at its minimum value (assuming the reaction is carried out at constant temperature and pressure). What this means is that the derivative of the Gibbs energy with respect to reaction coordinate (a measure of the extent of reaction that has occurred, ranging from zero for all reactants to a maximum for all products) vanishes, signalling a stationary point. This derivative is called the reaction Gibbs energy (or energy change) and corresponds to the difference between the chemical potentials of reactants and products at the composition of the reaction mixture. This criterion is both necessary and sufficient. If a mixture is not at equilibrium, the liberation of the excess Gibbs energy (or Helmholtz energy at constant volume reactions) is the "driving force" for the composition of the mixture to change until equilibrium is reached. The equilibrium constant can be related to the standard Gibbs free energy change for the reaction by the equation

$$\Delta_r G^\ominus = -RT \ln K_{eq}$$

where R is the universal gas constant and T the temperature.

When the reactants are dissolved in a medium of high ionic strength the quotient of activity coefficients may be taken to be constant. In that case the concentration quotient, K_c,

$$K_c = \frac{[S]^\sigma [T]^\tau}{[A]^\alpha [B]^\beta}$$

where [A] is the concentration of A, etc., is independent of the analytical concentration of the reactants. For this reason, equilibrium constants for solutions are usually determined in media of high ionic strength. K_c varies with ionic strength, temperature and pressure (or volume). Likewise K_p for gases depends on partial pressure. These constants are easier to measure and encountered in high-school chemistry courses.

Thermodynamics

At constant temperature and pressure, one must consider the Gibbs free energy, G, while at constant temperature and volume, one must consider the Helmholtz free energy: A, for the reaction; and at constant internal energy and volume, one must consider the entropy for the reaction: S.

The constant volume case is important in geochemistry and atmospheric chemistry where pressure variations are significant. Note that, if reactants and products were in standard state (completely pure), then there would be no reversibility and no equilibrium. Indeed, they would necessarily occupy disjoint volumes of space. The mixing of the products and reactants contributes a large entropy (known as entropy of mixing) to states containing equal mixture of products and reactants. The standard Gibbs energy change, together with the Gibbs energy of mixing, determine the equilibrium state.

In this article only the constant pressure case is considered. The relation between the Gibbs free energy and the equilibrium constant can be found by considering chemical potentials.

At constant temperature and pressure, the Gibbs free energy, G, for the reaction depends only on the extent of reaction: ξ, and can only decrease according to the second law of thermodynamics. It means that the derivative of G with ξ must be negative if the reaction happens; at the equilibrium the derivative being equal to zero.

$$\left(\frac{dG}{d\xi} \right)_{T,p} = 0: \quad \text{equilibrium}$$

In order to meet the thermodynamic condition for equilibrium, the Gibbs energy must be stationary, meaning that the derivative of G with respect to the extent of reaction: ξ, must be zero. It can be shown that in this case, the sum of chemical potentials of the products is equal to the sum of those corresponding to the reactants. Therefore, the sum of the Gibbs energies of the reactants must be the equal to the sum of the Gibbs energies of the products.

$$\alpha \mu_A + \beta \mu_B = \sigma \mu_S + \tau \mu_T$$

where μ is in this case a partial molar Gibbs energy, a chemical potential. The chemical potential of a reagent A is a function of the activity, {A} of that reagent.

$$\mu_A = \mu_A^\ominus + RT \ln\{A\}$$

(where $\mu\ominus$
A is the standard chemical potential).

The definition of the Gibbs energy equation interacts with the fundamental thermodynamic relation to produce

$$dG = Vdp - SdT + \sum_{i=1}^{k} \mu_i dN_i.$$

Inserting $dN_i = v_i\, d\xi$ into the above equation gives a Stoichiometric coefficient (v_i) and a differential that denotes the reaction occurring once ($d\xi$). At constant pressure and temperature the above equations can be written as

$$\left(\frac{dG}{d\xi}\right)_{T,p} = \sum_{i=1}^{k} \mu_i v_i = \Delta_r G_{T,p}$$ which is the "Gibbs free energy change for the reaction .

This results in:

$$\Delta_r G_{T,p} = \sigma\mu_S + \tau\mu_T - \alpha\mu_A - \beta\mu_B .$$

By substituting the chemical potentials:

$$\Delta_r G_{T,p} = (\sigma\mu_S^\ominus + \tau\mu_T^\ominus) - (\alpha\mu_A^\ominus + \beta\mu_B^\ominus) + (\sigma RT \ln\{S\} + \tau RT \ln\{T\}) - (\alpha RT \ln\{A\} + \beta RT \ln\{B\}),,$$

the relationship becomes:

$$\Delta_r G_{T,p} = \sum_{i=1}^{k} \mu_i^\ominus v_i + RT \ln \frac{\{S\}^\sigma \{T\}^\tau}{\{A\}^\alpha \{B\}^\beta}$$

$$\sum_{i=1}^{k} \mu_i^\ominus v_i = \Delta_r G^\ominus :$$

which is the standard Gibbs energy change for the reaction that can be calculated using thermodynamical tables. The reaction quotient is defined as:

$$Q_r = \frac{\{S\}^\sigma \{T\}^\tau}{\{A\}^\alpha \{B\}^\beta}$$

Therefore,

$$\left(\frac{dG}{d\xi}\right)_{T,p} = \Delta_r G_{T,p} = \Delta_r G^\ominus + RT \ln Q_r$$

At equilibrium:

$$\left(\frac{dG}{d\xi}\right)_{T,p} = \Delta_r G_{T,p} = 0$$

leading to:

$$0 = \Delta_r G^\ominus + RT \ln K_{eq}$$

and

$$\Delta_r G^\ominus = -RT \ln K_{eq}$$

Obtaining the value of the standard Gibbs energy change, allows the calculation of the equilibrium constant.

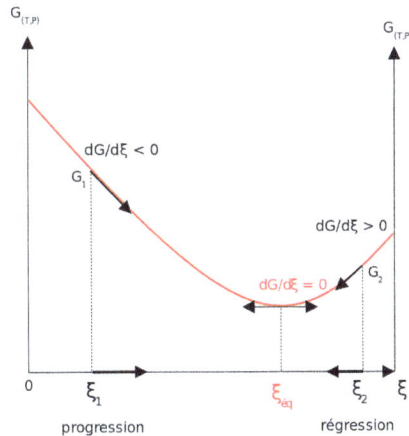

Diagramme G $_{(T,p)}$ = f (ξ)

Addition of Reactants or Products

For a reactional system at equilibrium: $Q_r = K_{eq}$; $\xi = \xi_{eq}$.

- If are modified activities of constituents, the value of the reaction quotient changes and becomes different from the equilibrium constant: $Q_r \neq K_{eq}$

$$\left(\frac{dG}{d\xi}\right)_{T,p} = \Delta_r G^\ominus + RT \ln Q_r$$

and

$$\Delta_r G^{\ominus} = -RT \ln K_{eq}$$

then

$$\left(\frac{dG}{d\xi}\right)_{T,p} = RT \ln\left(\frac{Q_r}{K_{eq}}\right)$$

- If activity of a reagent i increases

$$Q_r = \frac{\prod(a_j)^{v_j}}{\prod(a_i)^{v_i}} \text{ , the reaction quotient decreases.}$$

then

$$Q_r < K_{eq} \qquad \text{and} \qquad \left(\frac{dG}{d\xi}\right)_{T,p} < 0$$

The reaction will shift to the right (i.e. in the forward direction, and thus more products will form).

- If activity of a product j increases

then

$$Q_r > K_{eq} \qquad \text{and} \qquad \left(\frac{dG}{d\xi}\right)_{T,p} > 0$$

The reaction will shift to the left (i.e. in the reverse direction, and thus less products will form).

Note that activities and equilibrium constants are dimensionless numbers.

Treatment of Activity

The expression for the equilibrium constant can be rewritten as the product of a concentration quotient, K_c and an activity coefficient quotient, Γ.

$$K = \frac{[S]^{\sigma}[T]^{\tau}\cdots}{[A]^{\alpha}[B]^{\beta}\cdots} \times \frac{\gamma_S^{\sigma}\gamma_T^{\tau}\cdots}{\gamma_A^{\alpha}\gamma_B^{\beta}\cdots} = K_c\Gamma$$

[A] is the concentration of reagent A, etc. It is possible in principle to obtain values of the activity coefficients, γ. For solutions, equations such as the Debye–Hückel equation or extensions such as Davies equation Specific ion interaction theory or Pitzer equations may be used. However this is not always possible. It is common practice to assume that Γ is a constant, and to use the concentration quotient in place of the thermodynamic equilibrium constant. It is also general practice to use the term *equilibrium constant* instead of the more accurate *concentration quotient*. This practice will be followed here.

For reactions in the gas phase partial pressure is used in place of concentration and fugacity coefficient in place of activity coefficient. In the real world, for example, when making ammonia in industry, fugacity coefficients must be taken into account. Fugacity, f, is the product of partial pressure and fugacity coefficient. The chemical potential of a species in the gas phase is given by

$$\mu = \mu^{\ominus} + RT \ln\left(\frac{f}{\text{bar}}\right) = \mu^{\ominus} + RT \ln\left(\frac{p}{\text{bar}}\right) + RT \ln \gamma$$

so the general expression defining an equilibrium constant is valid for both solution and gas phases.

Concentration Quotients

In aqueous solution, equilibrium constants are usually determined in the presence of an "inert" electrolyte such as sodium nitrate $NaNO_3$ or potassium perchlorate $KClO_4$. The ionic strength of a solution is given by

$$I = \frac{1}{2} \sum_{i=1}^{N} c_i z_i^2$$

where c_i and z_i stand for the concentration and ionic charge of ion type i, and the sum is taken over all the N types of charged species in solution. When the concentration of dissolved salt is much higher than the analytical concentrations of the reagents, the ions originating from the dissolved salt determine the ionic strength, and the ionic strength is effectively constant. Since activity coefficients depend on ionic strength the activity coefficients of the species are effectively independent of concentration. Thus, the assumption that Γ is constant is justified. The concentration quotient is a simple multiple of the equilibrium constant.

$$K_c = \frac{K}{\Gamma}$$

However, K_c will vary with ionic strength. If it is measured at a series of different ionic strengths the value can be extrapolated to zero ionic strength. The concentration quotient obtained in this manner is known, paradoxically, as a thermodynamic equilibrium constant.

To use a published value of an equilibrium constant in conditions of ionic strength different from the conditions used in its determination, the value should be adjusted.

Metastable Mixtures

A mixture may be appear to have no tendency to change, though it is not at equilibrium. For example, a mixture of SO_2 and O_2 is metastable as there is a kinetic barrier to formation of the product, SO_3.

$$2 \, SO_2 + O_2 \rightleftharpoons 2 \, SO_3$$

The barrier can be overcome when a catalyst is also present in the mixture as in the contact process, but the catalyst does not affect the equilibrium concentrations.

Likewise, the formation of bicarbonate from carbon dioxide and water is very slow under normal conditions

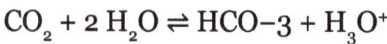

$$CO_2 + 2 \, H_2O \rightleftharpoons HCO-3 + H_3O^+$$

but almost instantaneous in the presence of the catalytic enzyme carbonic anhydrase.

Pure Substances

When pure substances (liquids or solids) are involved in equilibria their activities do not appear in the equilibrium constant because their numerical values are considered one.

Applying the general formula for an equilibrium constant to the specific case of a dilute solution of acetic acid in water one obtains

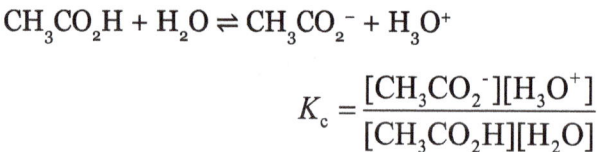

$$CH_3CO_2H + H_2O \rightleftharpoons CH_3CO_2^- + H_3O^+$$

$$K_c = \frac{[CH_3CO_2^-][H_3O^+]}{[CH_3CO_2H][H_2O]}$$

For all but very concentrated solutions, the water can be considered a "pure" liquid, and therefore it has an activity of one. The equilibrium constant expression is therefore usually written as

$$K = \frac{[CH_3CO_2^-][H_3O^+]}{[CH_3CO_2H]} = K_c.$$

A particular case is the self-ionization of water itself

$$2 \, H_2O \rightleftharpoons H_3O^+ + OH^-$$

Because water is the solvent, and has an activity of one, the self-ionization constant of water is defined as

$$K_w = [H^+][OH^-]$$

It is perfectly legitimate to write $[H^+]$ for the hydronium ion concentration, since the state of solvation of the proton is constant (in dilute solutions) and so does not affect the equilibrium concentrations. K_w varies with variation in ionic strength and/or temperature.

The concentrations of H^+ and OH^- are not independent quantities. Most commonly $[OH^-]$ is replaced by $K_w[H^+]^{-1}$ in equilibrium constant expressions which would otherwise include hydroxide ion.

Solids also do not appear in the equilibrium constant expression, if they are considered to be pure and thus their activities taken to be one. An example is the Boudouard reaction:

$$2\ CO \rightleftharpoons CO_2 + C$$

for which the equation (without solid carbon) is written as:

$$K_c = \frac{[CO_2]}{[CO]^2}$$

Multiple Equilibria

Consider the case of a dibasic acid H_2A. When dissolved in water, the mixture will contain H_2A, HA^- and A^{2-}. This equilibrium can be split into two steps in each of which one proton is liberated.

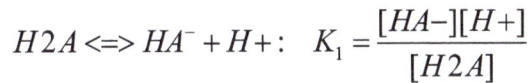

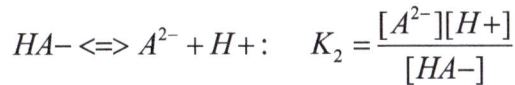

$$H2A <=> HA^- + H+: \quad K_1 = \frac{[HA-][H+]}{[H2A]}$$

$$HA- <=> A^{2-} + H+: \quad K_2 = \frac{[A^{2-}][H+]}{[HA-]}$$

K_1 and K_2 are examples of *stepwise* equilibrium constants. The *overall* equilibrium constant, β_D, is product of the stepwise constants.

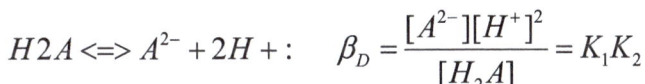

$$H2A <=> A^{2-} + 2H+: \quad \beta_D = \frac{[A^{2-}][H^+]^2}{[H_2A]} = K_1 K_2$$

Note that these constants are dissociation constants because the products on the right hand side of the equilibrium expression are dissociation products. In many systems, it is preferable to use association constants.

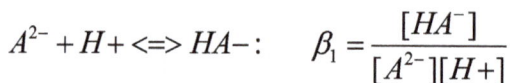

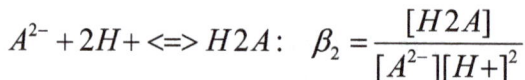

$$A^{2-} + H+ <=> HA-: \quad \beta_1 = \frac{[HA^-]}{[A^{2-}][H+]}$$

$$A^{2-} + 2H+ <=> H2A: \quad \beta_2 = \frac{[H2A]}{[A^{2-}][H+]^2}$$

β_1 and β_2 are examples of association constants. Clearly β_1 = $1/K_2$ and β_2 = $1/\beta_D$; $\log \beta_1$ = pK_2 and $\log \beta_2$ = $pK_2 + pK_1$ For multiple equilibrium systems.

Effect of Temperature

The effect of changing temperature on an equilibrium constant is given by the van 't Hoff equation

$$\frac{d \ln K}{dT} = \frac{\Delta H_m^\ominus}{RT^2}$$

Thus, for exothermic reactions (ΔH is negative), K decreases with an increase in temperature, but, for endothermic reactions, (ΔH is positive) K increases with an increase temperature. An alternative formulation is

$$\frac{d \ln K}{d(T^{-1})} = -\frac{\Delta H_m^\ominus}{R}$$

At first sight this appears to offer a means of obtaining the standard molar enthalpy of the reaction by studying the variation of K with temperature. In practice, however, the method is unreliable because error propagation almost always gives very large errors on the values calculated in this way.

Effect of Electric and Magnetic Fields

The effect of electric field on equilibrium has been studied by Manfred Eigen among others.

Types of Equilibrium

Haber–Bosch process

1. N_2 (g) $\rightleftharpoons$ N_2 (adsorbed)

2. N_2 (adsorbed) $\rightleftharpoons$ 2 N (adsorbed)

3. H_2 (g) $\rightleftharpoons$ H_2 (adsorbed)

4. H_2 (adsorbed) $\rightleftharpoons$ 2 H (adsorbed)

5. N (adsorbed) + 3 H(adsorbed) $\rightleftharpoons$ NH_3 (adsorbed)

6. NH_3 (adsorbed) $\rightleftharpoons$ NH_3 (g)

- In the gas phase: rocket engines

- The industrial synthesis such as ammonia in the Haber–Bosch process (depicted right) takes place through a succession of equilibrium steps including adsorption processes

- Atmospheric chemistry

- Seawater and other natural waters: chemical oceanography

- Distribution between two phases

    o   log *D* distribution coefficient: important for pharmaceuticals where lipophilicity is a significant property of a drug

    o   Liquid–liquid extraction, Ion exchange, Chromatography

    o   Solubility product

    o   Uptake and release of oxygen by haemoglobin in blood

- Acid–base equilibria: acid dissociation constant, hydrolysis, buffer solutions, indicators, acid–base homeostasis

- Metal–ligand complexation: sequestering agents, chelation therapy, MRI contrast reagents, Schlenk equilibrium

- Adduct formation: host–guest chemistry, supramolecular chemistry, molecular recognition, dinitrogen tetroxide

- In certain oscillating reactions, the approach to equilibrium is not asymptotically but in the form of a damped oscillation .

- The related Nernst equation in electrochemistry gives the difference in electrode potential as a function of redox concentrations.

- When molecules on each side of the equilibrium are able to further react irreversibly in secondary reactions, the final product ratio is determined according to the Curtin–Hammett principle.

In these applications, terms such as stability constant, formation constant, binding constant, affinity constant, association/dissociation constant are used. In biochemistry, it is common to give units for binding constants, which serve to define the concentration units used when the constant's value was determined.

Composition of a Mixture

When the only equilibrium is that of the formation of a 1:1 adduct as the composition of a mixture, there are any number of ways that the composition of a mixture can be calculated. For example, see ICE table for a traditional method of calculating the pH of a solution of a weak acid.

There are three approaches to the general calculation of the composition of a mixture at equilibrium.

1. The most basic approach is to manipulate the various equilibrium constants until the desired concentrations are expressed in terms of measured equilibrium constants (equivalent to measuring chemical potentials) and initial conditions.

2. Minimize the Gibbs energy of the system.

3. Satisfy the equation of mass balance. The equations of mass balance are simply statements that demonstrate that the total concentration of each reactant must be constant by the law of conservation of mass.

Mass-balance Equations

In general, the calculations are rather complicated or complex. For instance, in the case of a dibasic acid, H_2A dissolved in water the two reactants can be specified as the conjugate base, A^{2-}, and the proton, H^+. The following equations of mass-balance could apply equally well to a base such as 1,2-diaminoethane, in which case the base itself is designated as the reactant A:

$$T_A = [A] + [HA] + [H_2A]$$

$$T_H = [H] + [HA] + 2[H_2A] - [OH]$$

With T_A the total concentration of species A. Note that it is customary to omit the ionic charges when writing and using these equations.

When the equilibrium constants are known and the total concentrations are specified there are two equations in two unknown "free concentrations" [A] and [H]. This follows from the fact that $[HA] = \beta_1[A][H]$, $[H_2A] = \beta_2[A][H]^2$ and $[OH] = K_w[H]^{-1}$

$$T_A = [A] + \beta_1[A][H] + \beta_2[A][H]^2$$

$$T_H = [H] + \beta_1[A][H] + 2\beta_2[A][H]^2 - K_w[H]^{-1}$$

so the concentrations of the "complexes" are calculated from the free concentrations and the equilibrium constants. General expressions applicable to all systems with two reagents, A and B would be

$$T_A = [A] + \sum_i p_i \beta_i [A]^{p_i} [B]^{q_i}$$

$$T_B = [B] + \sum_i q_i \beta_i [A]^{p_i} [B]^{q_i}$$

It is easy to see how this can be extended to three or more reagents.

Polybasic Acids

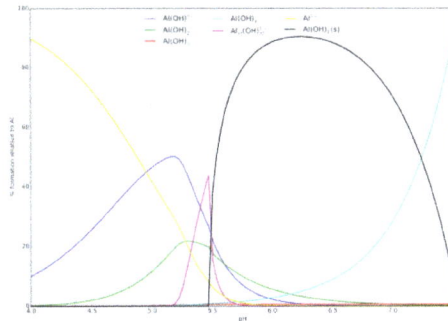

Species concentrations during hydrolysis of the aluminium.

The composition of solutions containing reactants A and H is easy to calculate as a function of p[H]. When [H] is known, the free concentration [A] is calculated from the mass-balance equation in A.

The diagram alongside, shows an example of the hydrolysis of the aluminium Lewis acid $Al^{3+}_{(aq)}$ shows the species concentrations for a 5×10^{-6} M solution of an aluminium salt as a function of pH. Each concentration is shown as a percentage of the total aluminium.

Solution and Precipitation

The diagram above illustrates the point that a precipitate that is not one of the main species in the solution equilibrium may be formed. At pH just below 5.5 the main species present in a 5 µM solution of Al^{3+} are aluminium hydroxides $Al(OH)^{2+}$, $AlOH+2$ and $Al13(OH)7+32$, but on raising the pH $Al(OH)_3$ precipitates from the solution. This occurs because $Al(OH)_3$ has a very large lattice energy. As the pH rises more and more $Al(OH)_3$ comes out of solution. This is an example of Le Châtelier's principle in action: Increasing the concentration of the hydroxide ion causes more aluminium hydroxide to precipitate, which removes hydroxide from the solution. When the hydroxide concentration becomes sufficiently high the soluble aluminate, $Al(OH)-4$, is formed.

Another common instance where precipitation occurs is when a metal cation interacts with an anionic ligand to form an electrically neutral complex. If the complex is hydrophobic, it will precipitate out of water. This occurs with the nickel ion Ni^{2+} and dimethylglyoxime, $(dmgH_2)$: in this case the lattice energy of the solid is not particularly large, but it greatly exceeds the energy of solvation of the molecule $Ni(dmgH)_2$.

Minimization of Free Energy

At equilibrium, G is at a minimum:

$$dG = \sum_{j=1}^{m} \mu_j \, dN_j = 0$$

For a closed system, no particles may enter or leave, although they may combine in various ways. The total number of atoms of each element will remain constant. This means that the minimization above must be subjected to the constraints:

$$\sum_{j=1}^{m} a_{ij} N_j = b_i^0$$

where a_{ij} is the number of atoms of element i in molecule j and bo i is the total number of atoms of element i, which is a constant, since the system is closed. If there are a total of k types of atoms in the system, then there will be k such equations.

This is a standard problem in optimisation, known as constrained minimisation. The most common method of solving it is using the method of Lagrange multipliers, also known as undetermined multipliers (though other methods may be used).

Define:

$$\mathcal{G} = G + \sum_{i=1}^{k} \lambda_i \left(\sum_{j=1}^{m} a_{ij} N_j - b_i^0 \right) = 0$$

where the λ_i are the Lagrange multipliers, one for each element. This allows each of the N_j to be treated independently, and it can be shown using the tools of multivariate calculus that the equilibrium condition is given by

$$\frac{\partial \mathcal{G}}{\partial N_j} = 0 \quad \text{and} \quad \frac{\partial \mathcal{G}}{\partial \lambda_i} = 0$$

This is a set of $(m + k)$ equations in $(m + k)$ unknowns (the N_j and the λ_i) and may, therefore, be solved for the equilibrium concentrations N_j as long as the chemical potentials are known as functions of the concentrations at the given temperature and pressure.

This method of calculating equilibrium chemical concentrations is useful for systems with a large number of different molecules. The use of k atomic element conservation equations for the mass constraint is straightforward, and replaces the use of the stoichiometric coefficient equations.

Unit Operation

In chemical engineering and related fields, a unit operation is a basic step in a process. Unit operations involve a physical change or chemical transformation such as separation, crystallization, evaporation, filtration, polymerization, isomerization, and other reactions. For example, in milk processing, homogenization, pasteurization, chilling, and packaging are each unit operations which are connected to create the overall process. A process may require many unit operations to obtain the desired product from the starting materials, or feedstocks.

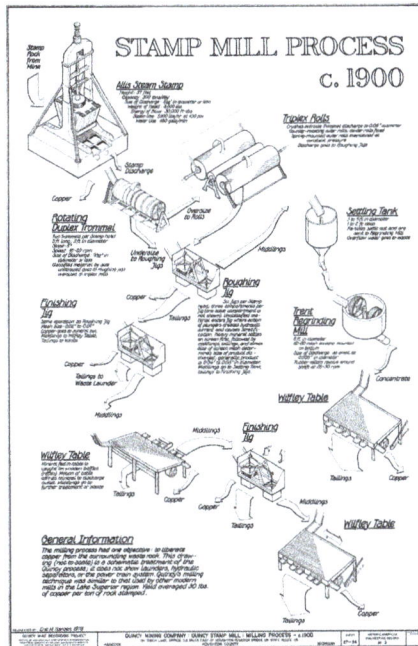

An ore extraction process broken into its constituent unit operations (Quincy Mine, Hancock, MI ca. 1900)

History

Historically, the different chemical industries were regarded as different industrial processes and with different principles. Arthur Dehon Little propounded the concept of "unit operations" to explain industrial chemistry processes in 1916. In 1923, William H. Walker, Warren K. Lewis and William H. McAdams wrote the book *The Principles of Chemical Engineering* and explained that the variety of chemical industries have processes which follow the same physical laws. They summed up these similar processes into unit operations. Each unit operation follows the same physical laws and may be used in all relevant chemical industries. For instance, the same engineering is required to design a mixer for either napalm or porridge, even if the use, market or manufacturers are very different. The unit operations form the fundamental principles of chemical engineering.

Chemical Engineering

Chemical engineering unit operations consist of five classes:

1. Fluid flow processes, including fluids transportation, filtration, and solids fluidization.

2. Heat transfer processes, including evaporation and heat exchange.

3. Mass transfer processes, including gas absorption, distillation, extraction, adsorption, and drying.

4. Thermodynamic processes, including gas liquefaction, and refrigeration.

5. Mechanical processes, including solids transportation, crushing and pulverization, and screening and sieving.

Chemical engineering unit operations also fall in the following categories which involve elements from more than one class:

- Combination (mixing)

- Separation (distillation, crystallization)

- Reaction (chemical reaction)

Furthermore, there are some unit operations which combine even these categories, such as reactive distillation and stirred tank reactors. A "pure" unit operation is a physical transport process, while a mixed chemical/physical process requires modeling both the physical transport, such as diffusion, *and* the chemical reaction. This is usually necessary for designing catalytic reactions, and is considered a separate discipline, termed chemical reaction engineering.

Chemical engineering unit operations and chemical engineering unit processing form the main principles of all kinds of chemical industries and are the foundation of designs of chemical plants, factories, and equipment used.

In general, unit operations are designed by writing down the balances for the transported quantity for each elementary component (which may be infinitesimal) in the form of equations, and solving the equations for the design parameters, then selecting an optimal solution out of the several possible and then designing the physical equipment. For instance, distillation in a plate column is analyzed by writing down the mass balances for each plate, wherein the known vapor-liquid equilibrium and efficiency, drip out and drip in comprise the total mass flows, with a sub-flow for each component. Combining a stack of these gives the system of equations for the whole column. There is a range of solutions, because a higher reflux ratio enables fewer plates, and vice versa. The engineer must then find the optimal solution with respect to acceptable volume holdup, column height and cost of construction.

Process Design

In Chemical engineering, process design is the design of processes for desired physical and/or chemical transformation of materials. Process design is central to chemical engineering, and it can be considered to be the summit of that field, bringing together all of the field's components.

Process design can be the design of new facilities or it can be the modification or expansion of existing facilities. The design starts at a conceptual level and ultimately ends in the form of fabrication and construction plans.

Process design is distinct from equipment design, which is closer in spirit to the design of unit operations. Processes often include many unit operations.

Documentation

Process design documents serve to define the design and they ensure that the design components fit together. They are useful in communicating ideas and plans to other engineers involved with the design, to external regulatory agencies, to equipment vendors and to construction contractors.

In order of increasing detail, process design documents include:

- Block flow diagrams (BFD): Very simple diagrams composed of rectangles and lines indicating major material or energy flows.

- Process flow diagrams (PFD): Typically more complex diagrams of major unit operations as well as flow lines. They usually include a material balance, and sometimes an energy balance, showing typical or design flowrates, stream compositions, and stream and equipment pressures and temperatures.

- Piping and instrumentation diagrams (P&ID): Diagrams showing each and every pipeline with piping class (carbon steel or stainless steel) and pipe size (diameter). They also show valving along with instrument locations and process control schemes.

- Specifications: Written design requirements of all major equipment items.

Process designers also typically write operating manuals on how to start-up, operate and shut-down the process.

Documents are maintained after construction of the process facility for the operating personnel to refer to. The documents also are useful when modifications to the facility are planned.

A primary method of developing the process documents is process flowsheeting.

Design Considerations

There are several considerations that need to be made when designing any chemical process unit. Design conceptualization and considerations can begin once product purities, yields, and throughput rates are all defined.

Objectives that a design may strive to include:

- Throughput rate
- Process yield
- Product purity

Constraints include:

- Capital cost
- Available space
- Safety concerns
- Environmental impact and projected effluents and emissions
- Waste production
- Operating and maintenance costs

Other factors that designers may include are:

- Reliability
- Redundancy
- Flexibility
- Anticipated variability in feedstock and allowable variability in product.

Sources of Design Information

Designers usually do not start from scratch, especially for complex projects. Often the engineers have pilot plant data available or data from full-scale operating facilities. Other sources of information include proprietary design criteria provided by process licensors, published scientific data, laboratory experiments, and input.

Computer Help

The advent of low cost powerful computers has aided complex mathematical simulation of processes, and simulation software is often used by design engineers. Simulations can identify weaknesses in designs and allow engineers to choose better alternatives.

However, engineers still rely on heuristics, intuition, and experience when designing a process. Human creativity is an element in complex designs.

Chemical Process Modeling

Chemical process modeling is a computer modeling technique used in chemical engineering process design. It typically involves using purpose-built software to define a system of interconnected components, which are then solved so that the steady-state or dynamic behavior of the system can be predicted. The system components and connections are represented as a Process Flow diagram. Simulations can be as simple as the mixing of two substances in a tank, or as complex as an entire alumina refinery.

Chemical process modeling requires a knowledge of the properties of the chemicals involved in the simulation, as well as the physical properties and characteristics of the components of the system, such as tanks, pumps, pipes, pressure vessels, and so on.

Process Integration

Process integration is a term in chemical engineering which has two possible meanings.

1. A holistic approach to process design which emphasizes the unity of the process and considers the interactions between different unit operations from the outset, rather than optimising them separately. This can also be called *integrated process design* or *process synthesis*. El-Halwagi (1997 and 2006) and Smith (2005) describe the approach well. An important first step is often *product design* (Cussler and Moggridge 2003) which develops the specification for the product to fulfil its required purpose.

2. *Pinch analysis*, a technique for designing a process to minimise energy consumption and maximise heat recovery, also known as *heat integration, energy integration* or *pinch technology*. The technique calculates thermodynamically attainable *energy targets* for a given process and identifies how to achieve them. A key insight is the pinch temperature, which is the most constrained point in the process. The most detailed explanation of the techniques is by Linnhoff et al. (1982), Shenoy (1995) and

Kemp (2006). This definition reflects the fact that the first major success for process integration was the thermal pinch analysis addressing energy problems and pioneered by Linnhoff and co-workers. Later, other pinch analyses were developed for several applications such as mass-exchange networks (El-Halwagi and Manousiouthakis, 1989), water minimization (Wang and Smith, 1994), and material recycle (El-Halwagi et al., 2003). A very successful extension was "Hydrogen Pinch", which was applied to refinery hydrogen management (Nick Hallale et al., 2002 and 2003). This allowed refiners to minimise the capital and operating costs of hydrogen supply to meet ever stricter environmental regulations and also increase hydrotreater yields.

In the context of chemical engineering, Process Integration can be defined as a holistic approach to process design and optimization, which exploits the interactions between different units in order to employ resources effectively and minimize costs.

Note that Process Integration is not limited to the design of new plants, but it also covers retrofit design (e.g. new units to be installed in an old plant) and the operation of existing systems. Nick Hallale (2001), in his article in Chemical Engineering Progress provided a state of the art review. He explained that process integration far wider scope and touches every area of process design. Industries are making more money from their raw materials and capital assets while becoming cleaner and more sustainable.

The main advantage of process integration is to consider a system as a whole (i.e. integrated or holistic approach) in order to improve their design and/or operation. In contrast, an analytical approach would attempt to improve or optimize process units separately without necessarily taking advantage of potential interactions among them.

For instance, by using process integration techniques it might be possible to identify that a process can use the heat rejected by another unit and reduce the overall energy consumption, even if the units are not running at optimum conditions on their own. Such an opportunity would be missed with an analytical approach, as it would seek to optimize each unit, and thereafter it wouldn't be possible to re-use the heat internally.

Typically, process integration techniques are employed at the beginning of a project (e.g. a new plant or the improvement of an existing one) to screen out promising options to optimize the design and/or operation of a process plant.

Also it is often employed, in conjunction with simulation and mathematical optimization tools to identify opportunities in order to better integrate a system (new or existing) and reduce capital and/or operating costs.

Most process integration techniques employ Pinch analysis or Pinch Tools to evaluate several processes as a whole system. Therefore, strictly speaking, both concepts are not the same, even if in certain contexts they are used interchangeably. The review by Nick Hallale (2001) explains that in the future, several trends are to be expected in the field.

In the future, it seems probable that the boundary between targets and design will be blurred and that these will be based on more structural information regarding the process network. Second, it is likely that we will see a much wider range of applications of process integration. There is still much work to be carried out in the area of separation, not only in complex distillation systems, but also in mixed types of separation systems. This includes processes involving solids, such as flotation and crystallization. The use of process integration techniques for reactor design has seen rapid progress, but is still in its early stages. Third, a new generation of software tools is expected. The emergence of commercial software for process integration is fundamental to its wider application in process design.

References

- Hudlický, Miloš (1996). Reductions in Organic Chemistry. Washington, D.C.: American Chemical Society. p. 429. ISBN 0-8412-3344-6.

- Robertson, William (2010). More Chemistry Basics. National Science Teachers Association. p. 82. ISBN 978-1-936137-74-9.

- Rodgers, Glen (2012). Descriptive Inorganic, Coordination, and Solid-State Chemistry. Brooks/Cole, Cengage Learning. p. 330. ISBN 978-0-8400-6846-0.

- March Jerry; (1985). Advanced Organic Chemistry reactions, mechanisms and structure (3rd ed.). New York: John Wiley & Sons, inc. ISBN 0-471-85472-7

- Stefanidakis, G.; Gwyn, J.E. (1993). "Alkylation". In John J. McKetta. Chemical Processing Handbook. CRC Press. pp. 80–138. ISBN 0-8247-8701-3.

- Stranges, Anthony N. (2000). "Germany's synthetic fuel industry, 1935–1940". In Lesch, John E. The German Chemical Industry in the Twentieth Century. Kluwer Academic Publishers. p. 170. ISBN 0-7923-6487-2.

- Kandori, Hideki (2006). "Retinal Binding Proteins". In Dugave, Christophe. Cis-trans Isomerization in Biochemistry. Wiley-VCH. p. 56. ISBN 3-527-31304-4.

- Glusker, Jenny P. (1991). "Structural Aspects of Metal Liganding to Functional Groups in Proteins". In Christian B. Anfinsen. Advances in Protein Chemistry. 42. San Diego: Academic Press. p. 7. ISBN 0-12-034242-1.

- Guo, Liang-Hong; Allen, H.; Hill, O. (1991). "Direct Electrochemistry of Proteins and Enzymes". In A. G. Sykes. Advances in Inorganic Chemistry. 36. San Diego: Academic Press. p. 359. ISBN 0-12-023636-2.

- Meyer, H. Jürgen (2007). "Festkörperchemie". In Erwin Riedel. Modern Inorganic Chemistry (in German) (3rd ed.). de Gruyter. p. 171. ISBN 978-3-11-019060-1.

- Elschenbroich, Christoph (2008). Organometallchemie (6th ed.). Wiesbaden: Vieweg+Teubner Verlag. p. 263. ISBN 978-3-8351-0167-8.

- March, Jerry (1985), Advanced Organic Chemistry: Reactions, Mechanisms, and Structure (3rd ed.), New York: Wiley, ISBN 0-471-85472-7

- Latscha, Hans Peter; Kazmaier, Uli; Klein, Helmut Alfons (2008). Organische Chemie: Chemie-basiswissen II (in German). 2 (6th ed.). Springer. p. 273. ISBN 978-3-540-77106-7.

Separation Process: Types and Techniques

Separation processes are methods that are used to convert mixtures of chemical substances into two or more distinct mixtures. The technical term used for the separation of mixtures is known as chromatography. The techniques elucidated in this section are of vital importance, and provides a better understanding of separation processes.

Separation Process

A separation is a method to achieve any phenomenon that converts a mixture of chemical substance into two or more distinct product mixtures, which may be referred to as mixture. at least one of which is enriched in one or more of the mixture's constituents. In some cases, a separation may fully divide the mixture into its pure constituents. Separations differences in chemical properties or physical properties such as size, shape, mass, density, or chemical affinity, between the constituents of a mixture. They are often classified according to the particular differences they use to achieve separation. Usually there is only physical movement and no substantial chemical modification. If no single difference can be used to accomplish a desired separation, multiple operations will often be performed in combination to achieve the desired end.

With a few exceptions, elements or compounds are naturally found in an impure state. Often these impure raw materials must be separated into their purified components before they can be put to productive use, making separation techniques essential for the modern industrial economy. In some cases, these separations require total purification, as in the electrolysis refining of bauxite ore for aluminum metal, but a good example of an incomplete separation technique is oil refining. Crude oil occurs naturally as a mixture of various hydrocarbons and impurities. The refining process splits this mixture into other, more valuable mixtures such as natural gas, gasoline and chemical feedstocks, none of which are pure substances, but each of which must be separated from the raw crude. In both of these cases, a series of separations is necessary to obtain the desired end products. In the case of oil refining, crude is subjected to a long series of individual distillation steps, each of which produces a different product or intermediate.

The purpose of a separation may be *analytical*, i.e. to help analyze components in the original mixture without any attempt to save the fractions, or may be *preparative*, i.e.

to "prepare" fractions or samples of the components that can be saved. The separation can be done on a small scale, effectively a laboratory scale for analytical or preparative purposes, or on a large scale, effectively an industrial scale for preparative purposes, or on some intermediate scale.

List of Separation Techniques

- Adsorption, adhesion of atoms, ions or molecules of gas, liquid, or dissolved solids to a surface

- Capillary electrophoresis

- Centrifugation and cyclonic separation, separates based on density differences

- Chromatography separates dissolved substances by different interaction with (i.e., travel through) a material

- Crystallization

- Decantation

- Demister (vapor), removes liquid droplets from gas streams

- Distillation, used for mixtures of liquids with different boiling points

- Drying, removes liquid from a solid by vaporisation

- Electrophoresis, separates organic molecules based on their different interaction with a gel under an electric potential (i.e., different travel)

- Electrostatic Separation, works on the principle of corona discharge, where two plates are placed close together and high voltage is applied. This high voltage is used to separate the ionized particles.

- Elutriation

- Evaporation

- Extraction

  - Leaching

  - Liquid-liquid extraction

  - Solid phase extraction

- Field flow fractionation

- Flotation

- o   Dissolved air flotation, removes suspended solids non-selectively from slurry by bubbles that are generated by air coming out of solution

- o   Froth flotation, recovers valuable, hydrophobic solids by attachment to air bubbles generated by mechanical agitation of an air-slurry mixture, which float, and are recovered

- o   Deinking, separating hydrophobic ink particles from hydrophilic paper pulp in paper recycling

- Flocculation, separates a solid from a liquid in a colloid, by use of a flocculant, which promotes the solid clumping into flocs

- Filtration – Mesh, bag and paper filters are used to remove large particulates suspended in fluids (e.g., fly ash) while membrane processes including microfiltration, ultrafiltration, nanofiltration, reverse osmosis, dialysis (biochemistry) utilising synthetic membranes, separates micrometre-sized or smaller species

- Fractional distillation

- Fractional freezing

- Oil-water separation, gravimetrically separates suspended oil droplets from waste water in oil refineries, petrochemical and chemical plants, natural gas processing plants and similar industries

- Magnetic separation

- Precipitation

- Recrystallization

- Scrubbing, separation of particulates (solids) or gases from a gas stream using liquid.

- Sedimentation, separates using vocal density pressure differences

  - o   Gravity separation

- Sieving

- Stripping

- Sublimation

- Vapor-liquid separation, separates by gravity, based on the Souders-Brown equation

- Winnowing

- Zone refining

Chromatography

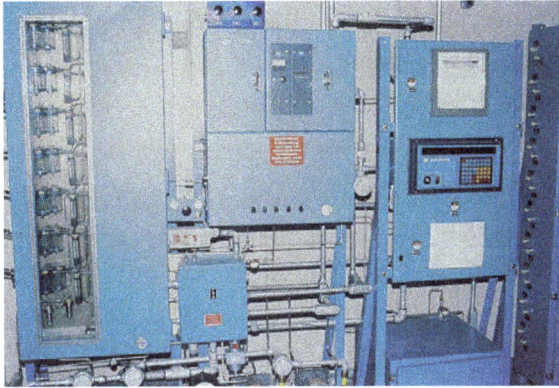

Pictured is a sophisticated gas chromatography system. This instrument records concentrations of acrylonitrile in the air at various points throughout the chemical laboratory.

Automated fraction collector and sampler for chromatographic techniques

Chromatography is the collective term for a set of laboratory techniques for the separation of mixtures. The mixture is dissolved in a fluid called the *mobile phase*, which carries it through a structure holding another material called the *stationary phase*. The various constituents of the mixture travel at different speeds, causing them to separate. The separation is based on differential partitioning between the mobile and stationary phases. Subtle differences in a compound's partition coefficient result in differential retention on the stationary phase and thus changing the separation.

Chromatography may be preparative or analytical. The purpose of preparative chromatography is to separate the components of a mixture for more advanced use (and is thus a form of purification). Analytical chromatography is done normally with smaller amounts of material and is for measuring the relative proportions of analytes in a mixture. The two are not mutually exclusive.

History

Chromatography was first employed in Russia by the Italian-born scientist Mikhail Tsvet in 1900. He continued to work with chromatography in the first decade of the 20th century, primarily for the separation of plant pigments such as chlorophyll, carotenes, and xanthophylls. Since these components have different colors (green, orange, and yellow, respectively) they gave the technique its name. New types of chromatography developed during the 1930s and 1940s made the technique useful for many separation processes.

Thin layer chromatography is used to separate components of a plant extract, illustrating the experiment with plant pigments that gave chromatography its name

Chromatography technique developed substantially as a result of the work of Archer John Porter Martin and Richard Laurence Millington Synge during the 1940s and 1950s, for which they won a Nobel prize. They established the principles and basic techniques of partition chromatography, and their work encouraged the rapid development of several chromatographic methods: paper chromatography, gas chromatography, and what would become known as high performance liquid chromatography. Since then, the technology has advanced rapidly. Researchers found that the main principles of Tsvet's chromatography could be applied in many different ways, resulting in the different varieties of chromatography described below. Advances are continually improving the technical performance of chromatography, allowing the separation of increasingly similar molecules.

Chromatography Terms

- The analyte is the substance to be separated during chromatography. It is also normally what is needed from the mixture.

- Analytical chromatography is used to determine the existence and possibly also the concentration of analyte(s) in a sample.

- A bonded phase is a stationary phase that is covalently bonded to the support particles or to the inside wall of the column tubing.

- A chromatogram is the visual output of the chromatograph. In the case of an optimal separation, different peaks or patterns on the chromatogram correspond to different components of the separated mixture.

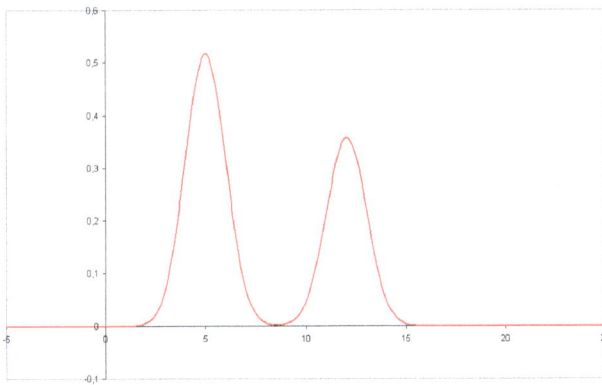

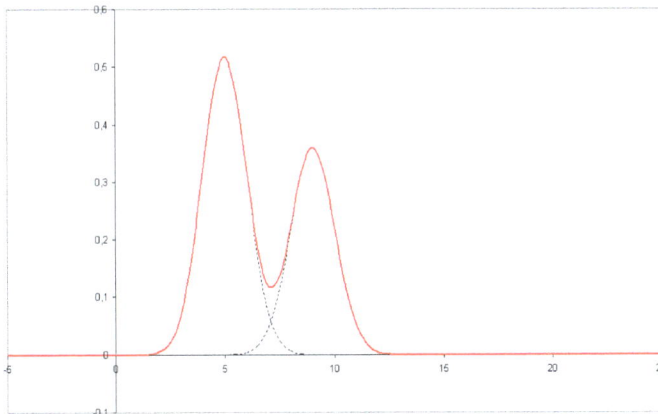

Plotted on the x-axis is the retention time and plotted on the y-axis a signal (for example obtained by a spectrophotometer, mass spectrometer or a variety of other detectors) corresponding to the response created by the analytes exiting the system. In the case of an optimal system the signal is proportional to the concentration of the specific analyte separated.

- A chromatograph is equipment that enables a sophisticated separation, e.g. gas chromatographic or liquid chromatographic separation.

- Chromatography is a physical method of separation that distributes components to separate between two phases, one stationary (stationary phase), the other (the mobile phase) moving in a definite direction.

- The eluate is the mobile phase leaving the column.

- The eluent is the solvent that carries the analyte.

- An eluotropic series is a list of solvents ranked according to their eluting power.

- An immobilized phase is a stationary phase that is immobilized on the support particles, or on the inner wall of the column tubing.

- The mobile phase is the phase that moves in a definite direction. It may be a liquid (LC and Capillary Electrochromatography (CEC)), a gas (GC), or a supercritical fluid (supercritical-fluid chromatography, SFC). The mobile phase consists of the sample being separated/analyzed and the solvent that moves the sample through the column. In the case of HPLC the mobile phase consists of a non-polar solvent(s) such as hexane in normal phase or a polar solvent such as methanol in reverse phase chromatography and the sample being separated. The mobile phase moves through the chromatography column (the stationary phase) where the sample interacts with the stationary phase and is separated.

- Preparative chromatography is used to purify sufficient quantities of a substance for further use, rather than analysis.

- The retention time is the characteristic time it takes for a particular analyte to pass through the system (from the column inlet to the detector) under set conditions.

- The sample is the matter analyzed in chromatography. It may consist of a single component or it may be a mixture of components. When the sample is treated in the course of an analysis, the phase or the phases containing the analytes of interest is/are referred to as the sample whereas everything out of interest separated from the sample before or in the course of the analysis is referred to as waste.

- The solute refers to the sample components in partition chromatography.

- The solvent refers to any substance capable of solubilizing another substance, and especially the liquid mobile phase in liquid chromatography.

- The stationary phase is the substance fixed in place for the chromatography procedure. Examples include the silica layer in thin layer chromatography

- The detector refers to the instrument used for qualitative and quantitative detection of analytes after separation.

Chromatography is based on the concept of partition coefficient. Any solute partitions between two immiscible solvents. When we make one solvent immobile (by adsorption on a solid support matrix) and another mobile it results in most common applications of chromatography. If the matrix support, or stationary phase, is polar (e.g. paper, silica etc.) it is forward phase chromatography, and if it is non-polar (C-18) it is reverse phase.

Techniques by Chromatographic Bed Shape

Column Chromatography

Column chromatography is a separation technique in which the stationary bed is within a tube. The particles of the solid stationary phase or the support coated with a liquid stationary phase may fill the whole inside volume of the tube (packed column) or be concentrated on or along the inside tube wall leaving an open, unrestricted path for the mobile phase in the middle part of the tube (open tubular column). Differences in rates of movement through the medium are calculated to different retention times of the sample.

In 1978, W. Clark Still introduced a modified version of column chromatography called flash column chromatography (flash). The technique is very similar to the traditional column chromatography, except for that the solvent is driven through the column by applying positive pressure. This allowed most separations to be performed in less than 20 minutes, with improved separations compared to the old method. Modern flash chromatography systems are sold as pre-packed plastic cartridges, and the solvent is pumped through the cartridge. Systems may also be linked with detectors and fraction collectors providing automation. The introduction of gradient pumps resulted in quicker separations and less solvent usage.

In expanded bed adsorption, a fluidized bed is used, rather than a solid phase made by a packed bed. This allows omission of initial clearing steps such as centrifugation and filtration, for culture broths or slurries of broken cells.

Phosphocellulose chromatography utilizes the binding affinity of many DNA-binding proteins for phosphocellulose. The stronger a protein's interaction with DNA, the higher the salt concentration needed to elute that protein.

Planar Chromatography

Planar chromatography is a separation technique in which the stationary phase is present as or on a plane. The plane can be a paper, serving as such or impregnated by a substance as the stationary bed (paper chromatography) or a layer of solid particles spread on a support such as a glass plate (thin layer chromatography). Different compounds in the sample mixture travel different distances according to how strongly they interact with the stationary phase as compared to the mobile phase. The specific Retention factor (R_f) of each chemical can be used to aid in the identification of an unknown substance.

Paper Chromatography

Paper chromatography is a technique that involves placing a small dot or line of sample solution onto a strip of *chromatography paper*. The paper is placed in a container with a shallow layer of solvent and sealed. As the solvent rises through the paper, it meets the sample mixture, which starts to travel up the paper with the solvent. This paper is made of cellulose, a polar substance, and the compounds within the mixture travel farther if they are non-polar. More polar substances bond with the cellulose paper more quickly, and therefore do not travel as far.

Thin Layer Chromatography (TLC)

Thin layer chromatography (TLC) is a widely employed laboratory technique and is similar to paper chromatography. However, instead of using a stationary phase of paper, it involves a stationary phase of a thin layer of adsorbent like silica gel, alumina, or cellulose on a flat, inert substrate. Compared to paper, it has the advantage of faster runs, better separations, and the choice between different adsorbents. For even better resolution and to allow for quantification, high-performance TLC can be used. An older popular use had been to differentiate chromosomes by observing distance in gel (separation of was a separate step).

Displacement Chromatography

The basic principle of displacement chromatography is: A molecule with a high affinity for the chromatography matrix (the displacer) competes effectively for binding sites, and thus displace all molecules with lesser affinities. There are distinct differences between displacement and elution chromatography. In elution mode, substances typically emerge from a column in narrow, Gaussian peaks. Wide separation of peaks, preferably to baseline, is desired for maximum purification. The speed at which any component of a mixture travels down the column in elution mode depends on many factors. But for two substances to travel at different speeds, and thereby be resolved, there must be substantial differences in some interaction between the biomolecules and the chromatography matrix. Operating parameters are adjusted to maximize the effect of this difference. In many cases, baseline separation of the peaks can be achieved only with gradient elution and low column loadings. Thus, two drawbacks to elution mode chromatography, especially at the preparative scale, are operational complexity, due to gradient solvent pumping, and low throughput, due to low column loadings. Displacement chromatography has advantages over elution chromatography in that components are resolved into consecutive zones of pure substances rather than "peaks". Because the process takes advantage of the nonlinearity of the isotherms, a larger column feed can be separated on a given column with the purified components recovered at significantly higher concentrations.

Techniques by Physical State of Mobile Phase

Gas Chromatography

Gas chromatography (GC), also sometimes known as gas-liquid chromatography, (GLC), is a separation technique in which the mobile phase is a gas. Gas chromatographic separation is always carried out in a column, which is typically "packed" or "capillary". Packed columns are the routine work horses of gas chromatography, being cheaper and easier to use and often giving adequate performance. Capillary columns generally give far superior resolution and although more expensive are becoming widely used, especially for complex mixtures. Both types of column are made from non-adsorbent and chemically inert materials. Stainless steel and glass are the usual materials for packed columns and quartz or fused silica for capillary columns.

Gas chromatography is based on a partition equilibrium of analyte between a solid or viscous liquid stationary phase (often a liquid silicone-based material) and a mobile gas (most often helium). The stationary phase is adhered to the inside of a small-diameter (commonly 0.53 – 0.18mm inside diameter) glass or fused-silica tube (a capillary column) or a solid matrix inside a larger metal tube (a packed column). It is widely used in analytical chemistry; though the high temperatures used in GC make it unsuitable for high molecular weight biopolymers or proteins (heat denatures them), frequently encountered in biochemistry, it is well suited for use in the petrochemical, environmental monitoring and remediation, and industrial chemical fields. It is also used extensively in chemistry research.

Liquid Chromatography

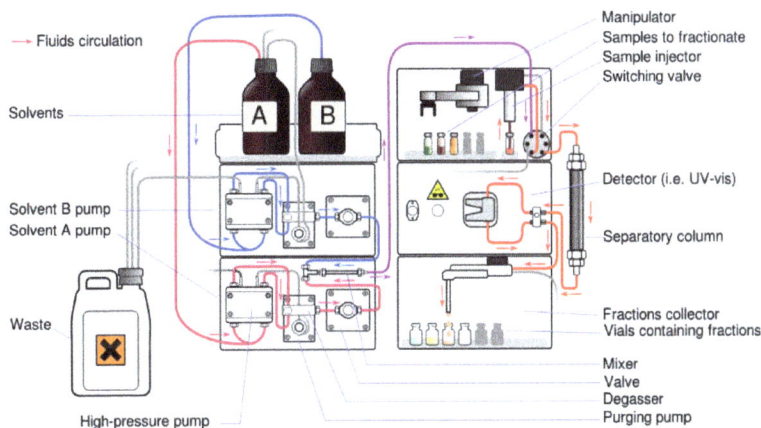

Preparative HPLC apparatus

Liquid chromatography (LC) is a separation technique in which the mobile phase is a liquid. It can be carried out either in a column or a plane. Present day liquid chromatography that generally utilizes very small packing particles and a relatively high pressure is referred to as high performance liquid chromatography (HPLC).

In HPLC the sample is forced by a liquid at high pressure (the mobile phase) through a column that is packed with a stationary phase composed of irregularly or spherically shaped particles, a porous monolithic layer, or a porous membrane. HPLC is historically divided into two different sub-classes based on the polarity of the mobile and stationary phases. Methods in which the stationary phase is more polar than the mobile phase (e.g., toluene as the mobile phase, silica as the stationary phase) are termed normal phase liquid chromatography (NPLC) and the opposite (e.g., water-methanol mixture as the mobile phase and C18 = octadecylsilyl as the stationary phase) is termed reversed phase liquid chromatography (RPLC).

Specific techniques under this broad heading are listed below.

Affinity Chromatography

Affinity chromatography is based on selective non-covalent interaction between an analyte and specific molecules. It is very specific, but not very robust. It is often used in biochemistry in the purification of proteins bound to tags. These fusion proteins are labeled with compounds such as His-tags, biotin or antigens, which bind to the stationary phase specifically. After purification, some of these tags are usually removed and the pure protein is obtained.

Affinity chromatography often utilizes a biomolecule's affinity for a metal (Zn, Cu, Fe, etc.). Columns are often manually prepared. Traditional affinity columns are used as a preparative step to flush out unwanted biomolecules.

However, HPLC techniques exist that do utilize affinity chromatography properties. Immobilized Metal Affinity Chromatography (IMAC) is useful to separate aforementioned molecules based on the relative affinity for the metal (I.e. Dionex IMAC). Often these columns can be loaded with different metals to create a column with a targeted affinity.

Supercritical Fluid Chromatography

Supercritical fluid chromatography is a separation technique in which the mobile phase is a fluid above and relatively close to its critical temperature and pressure.

Techniques by Separation Mechanism

Ion Exchange Chromatography

Ion exchange chromatography (usually referred to as ion chromatography) uses an ion exchange mechanism to separate analytes based on their respective charges. It is usually performed in columns but can also be useful in planar mode. Ion exchange chromatography uses a charged stationary phase to separate charged compounds including anions, cations, amino acids, peptides, and proteins. In conventional methods the stationary phase is an ion exchange resin that carries charged functional groups that interact with oppositely charged groups of the compound to retain. Ion exchange chromatography is commonly used to purify proteins using FPLC.

Size-exclusion Chromatography

Size-exclusion chromatography (SEC) is also known as gel permeation chromatography (GPC) or gel filtration chromatography and separates molecules according to their size (or more accurately according to their hydrodynamic diameter or hydrodynamic volume). Smaller molecules are able to enter the pores of the media and, therefore, molecules are trapped and removed from the flow of the mobile phase. The average residence time in the pores depends upon the effective size of the analyte molecules. However, molecules that are larger than the average pore size of the packing are excluded and thus suffer essentially no retention; such species are the first to be eluted. It is generally a low-resolution chromatography technique and thus it is often reserved for the final, "polishing" step of a purification. It is also useful for determining the tertiary structure and quaternary structure of purified proteins, especially since it can be carried out under native solution conditions.

Expanded Bed Adsorption Chromatographic Separation

An expanded bed chromatographic adsorption (EBA) column for a biochemical separation process comprises a pressure equalization liquid distributor having a self-cleaning function below a porous blocking sieve plate at the bottom of the expanded bed, an upper part nozzle assembly having a backflush cleaning function at the top of the expanded bed, a better distribution of the feedstock liquor added into the expanded bed ensuring that the fluid passed through the expanded bed layer displays a state of piston flow. The expanded bed layer displays a state of piston flow. The expanded bed chromatographic separation column has advantages of increasing the separation efficiency of the expanded bed.

Expanded-bed adsorption (EBA) chromatography is a convenient and effective technique for the capture of proteins directly from unclarified crude sample. In EBA chromatography, the settled bed is first expanded by upward flow of equilibration buffer. The crude feed, a mixture of soluble proteins, contaminants, cells, and cell debris, is then passed upward through the expanded bed. Target proteins are captured on the adsorbent, while particulates and contaminants pass through. A change to elution buffer while maintaining upward flow results in desorption of the target protein in expanded-bed mode. Alternatively, if the flow is reversed, the adsorbed particles will quickly settle and the proteins can be desorbed by an elution buffer. The mode used for elution (expanded-bed versus settled-bed) depends on the characteristics of the feed. After elution, the adsorbent is cleaned with a predefined cleaning-in-place (CIP) solution, with cleaning followed by either column regeneration (for further use) or storage.

Special Techniques

Reversed-phase Chromatography

Reversed-phase chromatography (RPC) is any liquid chromatography procedure in which the mobile phase is significantly more polar than the stationary phase. It is so named because in normal-phase liquid chromatography, the mobile phase is significantly less polar than the stationary phase. Hydrophobic molecules in the mobile phase tend to adsorb to the relatively hydrophobic stationary phase. Hydrophilic molecules in the mobile phase will tend to elute first. Separating columns typically comprise a C8 or C18 carbon-chain bonded to a silica particle substrate.

Hydrophobic Interaction Chromatography

Hydrophobic interactions between proteins and the chromatographic matrix can be exploited to purify proteins. In hydrophobic interaction chromatography the matrix material is lightly substituted with hydrophobic groups. These groups can range from methyl, ethyl, propyl, octyl, or phenyl groups.[] At high salt concentrations, non-polar sidechains on the surface on proteins "interact" with the hydrophobic groups; that is, both types of groups are excluded by the polar solvent (hydrophobic effects are augmented by increased ionic strength). Thus, the sample is applied to the column in a buffer which is highly polar. The eluant is typically an aqueous buffer with decreasing salt concentrations, increasing concentrations of detergent (which disrupts hydrophobic interactions), or changes in pH.

In general, Hydrophobic Interaction Chromatography (HIC) is advantageous if the sample is sensitive to pH change or harsh solvents typically used in other types of chromatography but not high salt concentrations. Commonly, it is the amount of salt in the buffer which is varied. In 2012, Müller and Franzreb described the effects of temperature on HIC using Bovine Serum Albumin (BSA) with four different types of hydrophobic resin. The study altered temperature as to effect the binding affinity of BSA onto the matrix. It was concluded that cycling temperature from 50 degrees to 10 degrees would not be adequate to effectively wash all BSA from the matrix but could be very effective if the column would only be used a few times. Using temperature to effect change allows labs to cut costs on buying salt and saves money.

If high salt concentrations along with temperature fluctuations want to be avoided you can use a more hydrophobic to compete with your sample to elute it. [source] This so-called salt independent method of HIC showed a direct isolation of Human Immunoglobulin G (IgG) from serum with satisfactory yield and used Beta-cyclodextrin as a competitor to displace IgG from the matrix. This largely opens up the possibility of using HIC with samples which are salt sensitive as we know high salt concentrations precipitate proteins.

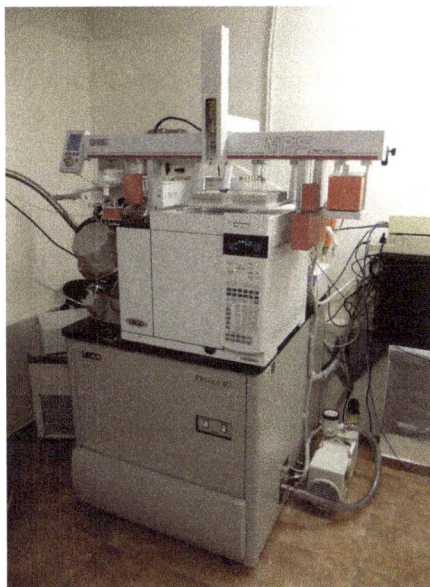

Two-dimensional chromatograph GCxGC-TOFMS at Chemical Faculty of GUT Gdańsk, Poland, 2016

Two-dimensional Chromatography

In some cases, the chemistry within a given column can be insufficient to separate some analytes. It is possible to direct a series of unresolved peaks onto a second column with different physico-chemical (Chemical classification) properties. Since the mechanism of retention on this new solid support is different from the first dimensional separation, it can be possible to separate compounds that are indistinguishable by one-dimensional chromatography. The sample is spotted at one corner of a square plate,developed, air-dried, then rotated by 90° and usually redeveloped in a second solvent system.

Simulated Moving-bed Chromatography

The simulated moving bed (SMB) technique is a variant of high performance liquid chromatography; it is used to separate particles and/or chemical compounds that would be difficult or impossible to resolve otherwise. This increased separation is brought about by a valve-and-column arrangement that is used to lengthen the stationary phase indefinitely. In the moving bed technique of preparative chromatography the feed entry and the analyte recovery are simultaneous and continuous, but because of practical difficulties with a continuously moving bed, simulated moving bed technique was proposed. In the simulated moving bed technique instead of moving the bed, the sample inlet and the analyte exit positions are moved continuously, giving the impression of a moving bed. True moving bed chromatography (TMBC) is only a theoretical concept. Its simulation, SMBC is achieved by the use of a multiplicity of columns in series and a complex valve arrangement, which provides for sample and solvent feed, and also analyte and waste takeoff at appropriate locations of any column, whereby it allows switching at regular intervals the sample entry in one direction, the solvent entry in the opposite direction, whilst changing the analyte and waste takeoff positions appropriately as well.

Pyrolysis Gas Chromatography

Pyrolysis gas chromatography mass spectrometry is a method of chemical analysis in which the sample is heated to decomposition to produce smaller molecules that are separated by gas chromatography and detected using mass spectrometry.

Pyrolysis is the thermal decomposition of materials in an inert atmosphere or a vacuum. The sample is put into direct contact with a platinum wire, or placed in a quartz sample tube, and rapidly heated to 600–1000 °C. Depending on the application even higher temperatures are used. Three different heating techniques are used in actual pyrolyzers: Isothermal furnace, inductive heating (Curie Point filament), and resistive heating using platinum filaments. Large molecules cleave at their weakest points and produce smaller, more volatile fragments. These fragments can be separated by gas chromatography. Pyrolysis GC chromatograms are typically complex because a wide range of different decomposition products is formed. The data can either be used as fingerprint to prove material identity or the GC/MS data is used to identify individual fragments to obtain structural information. To increase the volatility of polar fragments, various methylating reagents can be added to a sample before pyrolysis.

Besides the usage of dedicated pyrolyzers, pyrolysis GC of solid and liquid samples can be performed directly inside Programmable Temperature Vaporizer (PTV) injectors that provide quick heating (up to 30 °C/s) and high maximum temperatures of 600–650 °C. This is sufficient for some pyrolysis applications. The main advantage is that no dedicated instrument has to be purchased and pyrolysis can be performed as part of routine GC analysis. In this case quartz GC inlet liners have to be used. Quantitative data can be acquired, and good results of derivatization inside the PTV injector are published as well.

Fast Protein Liquid Chromatography

Fast protein liquid chromatography (FPLC), is a form of liquid chromatography that is often used to analyze or purify mixtures of proteins. As in other forms of chromatography, separation is possible because the different components of a mixture have different affinities for two materials, a moving fluid (the "mobile phase") and a porous solid (the "stationary phase"). In FPLC the mobile phase is an aqueous solution, or "buffer". The buffer flow rate is controlled by a positive-displacement pump and is normally kept constant, while the composition of the buffer can be varied by drawing fluids in different proportions from two or more external reservoirs. The stationary phase is a resin composed of beads, usually of cross-linked agarose, packed into a cylindrical glass or plastic column. FPLC resins are available in a wide range of bead sizes and surface ligands depending on the application.

Countercurrent Chromatography

Countercurrent chromatography (CCC) is a type of liquid-liquid chromatography, where both the stationary and mobile phases are liquids. The operating principle of CCC equipment requires a column consisting of an open tube coiled around a bobbin. The bobbin is rotated in a double-axis gyratory motion (a cardioid), which causes a variable gravity (G) field to act on the column during each rotation. This motion causes the column to see one partitioning step per revolution and components of the sample separate in the column due to their partitioning coefficient between the two immiscible liquid phases used. There are many types of CCC available today. These include HSCCC (High Speed CCC) and HPCCC (High Performance CCC). HPCCC is the latest and best performing version of the instrumentation available currently.

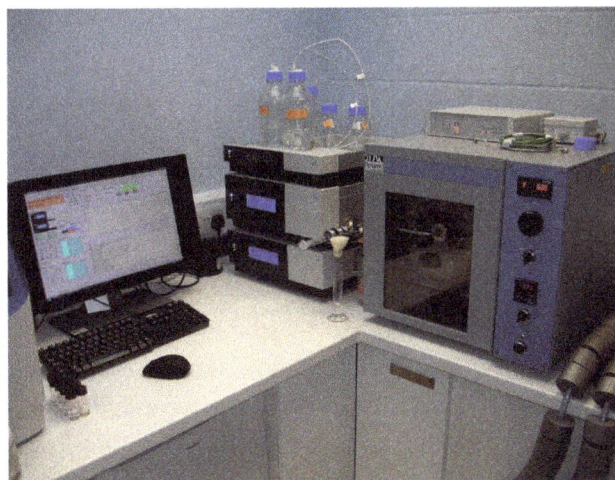

An example of a HPCCC system

Chiral Chromatography

Chiral chromatography involves the separation of stereoisomers. In the case of enantiomers, these have no chemical or physical differences apart from being three-dimensional mirror images. Conventional chromatography or other separation processes are incapable of separating them. To enable chiral separations to take place, either the mobile phase or the stationary phase must themselves be made chiral, giving differing affinities between the analytes. Chiral chromatography HPLC columns (with a chiral stationary phase) in both normal and reversed phase are commercially available.

Crystallization

Crystallization is the (natural or artificial) process where a solid forms where the atoms or molecules are highly organized in a structure known as a crystal. Some of the ways which crystals form are through precipitating from a solution, melt or more rarely deposited directly from a gas. Crystallization is also a chemical solid–liquid separation technique, in which mass transfer of a solute from the liquid solution to a pure solid crystalline phase occurs. In chemical engineering crystallization occurs in a crystallizer. Crystallization is therefore related to precipitation, although the result is not amorphous or disordered, but a crystal.

Process

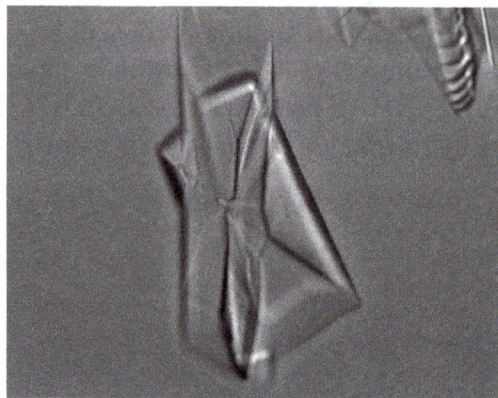

Time-lapse of growth of a citric acid crystal. The video covers an area of 2.0 by 1.5 mm and was captured over 7.2 min.

The crystallization process consists of two major events, *nucleation* and *crystal growth* which are driven by thermodynamic properties as well as chemical properties. In crystallization *Nucleation* is the step where the solute molecules or atoms dispersed in the solvent start to gather into clusters, on the microscopic scale (elevating solute concentration in a small region), that become stable under the current operating conditions. These stable clusters constitute the nuclei. Therefore, the clusters need to reach a critical size in order to become stable nuclei. Such critical size is dictated by many different factors (temperature, supersaturation, etc.). It is at the stage of nucleation that the atoms or molecules arrange in a defined and periodic manner that defines the crystal structure — note that "crystal structure" is a special term that refers to the relative arrangement of the atoms or molecules, not the macroscopic properties of the crystal (size and shape), although those are a result of the internal crystal structure.

The *crystal growth* is the subsequent size increase of the nuclei that succeed in achieving the critical cluster size. Crystal growth is a dynamic process occurring in equilibrium where solute molecules or atoms precipitate out of solution, and dissolve back into solution. Supersaturation is one of the driving forces of crystallization, as the solubility of a species is an equilibrium process quantified by Ksp. Depending upon the conditions, either nucleation or growth may be predominant over the other, dictating crystal size.

Many compounds have the ability to crystallize with some having different crystal structures, a phenomenon called polymorphism. Each polymorph is in fact a different thermodynamic solid state and crystal polymorphs of the same compound exhibit different physical properties, such as dissolution rate, shape (angles between facets and facet growth rates), melting point, etc. For this reason, polymorphism is of major importance in industrial manufacture of crystalline products. Additionally, crystal phases can sometimes be interconverted by varying factors such as temperature.

Crystallization in Nature

Snowflakes are a very well-known example, where subtle differences in *crystal growth* conditions result in different geometries.

Crystallized honey

There are many examples of natural process that involve crystallization.

Geological time scale process examples include:

- Natural (mineral) crystal formation;

- Stalactite/stalagmite, rings formation.

Usual time scale process examples include:

- Snow flakes formation;

- Honey crystallization (nearly all types of honey crystallize).

Methods

Crystal formation can be divided into two types, where the first type of crystals are composed of a cation and anion, also known as a salt, such as sodium acetate. The second type of crystals are composed of uncharged species, for example menthol.

Crystal formation can be achieved by various methods, such as: cooling, evaporation, addition of a second solvent to reduce the solubility of the solute (technique known as antisolvent or drown-out), solvent layering, sublimation, changing the cation or anion, as well as other methods.

The formation of a supersaturated solution does not guarantee crystal formation, and often a seed crystal or scratching the glass is required to form nucleation sites.

A typical laboratory technique for crystal formation is to dissolve the solid in a solution in which it is partially soluble, usually at high temperatures to obtain supersaturation. The hot mixture is then filtered to remove any insoluble impurities. The filtrate is allowed to slowly cool. Crystals that form are then filtered and washed with a solvent in which they are not soluble, but is miscible with the mother liquor. The process is then repeated to increase the purity in a technique known as recrystallization.

Typical Equipment

Equipment for the main industrial processes for crystallization.

1. *Tank crystallizers.* Tank crystallization is an old method still used in some specialized cases. Saturated solutions, in tank crystallization, are allowed to cool in open tanks. After a period of time the mother liquor is drained and the crystals removed. Nucleation and size of crystals are difficult to control. Typically, labor costs are very high.

Thermodynamic View

The nature of a crystallization process is governed by both thermodynamic and kinetic factors, which can make it highly variable and difficult to control. Factors such as impurity level, mixing regime, vessel design, and cooling profile can have a major impact on the size, number, and shape of crystals produced.

Low-temperature SEM magnification series for a snow crystal. The crystals are captured, stored, and sputter coated with platinum at cryo-temperatures for imaging.

Now put yourself in the place of a molecule within a pure and *perfect crystal*, being heated by an external source. At some sharply defined temperature, the complicated architecture of the crystal collapses to that of a liquid. Textbook thermodynamics says that melting occurs because the entropy, S, gain in your system by spatial randomization of the molecules has overcome the enthalpy, H, loss due to breaking the crystal packing forces:

$$T(S_{liquid} - S_{solid}) > H_{liquid} - H_{solid}$$

$$G_{liquid} < G_{solid}$$

This rule suffers no exceptions when the temperature is rising. By the same token, on cooling the melt, at the very same temperature the bell should ring again, and molecules should click back into the very same crystalline form. The entropy decrease due to the ordering of molecules within the system is overcompensated by the thermal randomization of the surroundings, due to the release of the heat of fusion; the entropy of the universe increases.

But liquids that behave in this way on cooling are the exception rather than the rule; in spite of the second principle of thermodynamics, crystallization usually occurs at lower temperatures (supercooling). This can only mean that a crystal is more easily destroyed than it is formed. Similarly, it is usually much easier to dissolve a perfect crystal in a solvent than to grow again a good crystal from the resulting solution. The nucleation and growth of a crystal are under kinetic, rather than thermodynamic, control.

Crystallization Dynamics

As mentioned above, a crystal is formed following a well-defined pattern, or structure, dictated by forces acting at the molecular level. As a consequence, during its formation process the crystal is in an environment where the solute concentration reaches a certain critical value, before changing status. Solid formation, impossible below the solubility threshold at the given temperature and pressure conditions, may then take place at a concentration higher than the theoretical solubility level. The difference between the actual value of the solute concentration at the crystallization limit and the theoretical (static) solubility threshold is called supersaturation and is a fundamental factor in crystallization.

Nucleation

Nucleation is the initiation of a phase change in a small region, such as the formation of a solid crystal from a liquid solution. It is a consequence of rapid local fluctuations on a molecular scale in a homogeneous phase that is in a state of metastable equilibrium. Total nucleation is the sum effect of two categories of nucleation – primary and secondary.

Primary Nucleation

Primary nucleation is the initial formation of a crystal where there are no other crystals present or where, if there are crystals present in the system, they do not have any influence on the process. This can occur in two conditions. The first is homogeneous nucleation, which is nucleation that is not influenced in any way by solids. These solids include the walls of the crystallizer vessel and particles of any foreign substance. The second category, then, is heterogeneous nucleation. This occurs when solid particles of foreign substances cause an increase in the rate of nucleation that would otherwise not be seen without the existence of these foreign particles. Homogeneous nucleation rarely occurs in practice due to the high energy necessary to begin nucleation without a solid surface to catalyse the nucleation.

Primary nucleation (both homogeneous and heterogeneous) has been modelled with the following:

$$B = \frac{dN}{dt} = k_n (c - c^*)^n$$

- B is the number of nuclei formed per unit volume per unit time.

- N is the number of nuclei per unit volume.

- k_n is a rate constant.

- c is the instantaneous solute concentration.

- c^* is the solute concentration at saturation.

- $(c - c^*)$ is also known as supersaturation.

- n is an empirical exponent that can be as large as 10, but generally ranges between 3 and 4.

Secondary Nucleation

Secondary nucleation is the formation of nuclei attributable to the influence of the existing microscopic crystals in the magma. The first type of known secondary crystallization is attributable to fluid shear, the other due to collisions between already existing crystals with either a solid surface of the crystallizer or with other crystals themselves. Fluid shear nucleation occurs when liquid travels across a Crystal at a high speed, sweeping away nuclei that would otherwise be incorporated into a Crystal, causing the swept-away nuclei to become new crystals. Contact nucleation has been found to be the most effective and common method for nucleation. The benefits include the following

- Low kinetic order and rate-proportional to supersaturation, allowing easy control without unstable operation.

- Occurs at low supersaturation, where growth rate is optimum for good quality.

- Low necessary energy at which crystals strike avoids the breaking of existing crystals into new crystals.

- The quantitative fundamentals have already been isolated and are being incorporated into practice.

Crystal growth

The following model, although somewhat simplified, is often used to model secondary nucleation:

- k_1 is a rate constant.

- M_T is the suspension density.

- • j is an empirical exponent that can range up to 1.5, but is generally 1.

- • b is an empirical exponent that can range up to 5, but is generally 2.

Crystal Growth

Once the first small crystal, the nucleus, forms it acts as a convergence point (if unstable due to supersaturation) for molecules of solute touching – or adjacent to – the crystal so that it increases its own dimension in successive layers. The pattern of growth resembles the rings of an onion, as shown in the picture, where each colour indicates the same mass of solute; this mass creates increasingly thin layers due to the increasing surface area of the growing crystal. The supersaturated solute mass the original nucleus may *capture* in a time unit is called the *growth rate* expressed in kg/(m²*h), and is a constant specific to the process. Growth rate is influenced by several physical factors, such as surface tension of solution, pressure, temperature, relative crystal velocity in the solution, Reynolds number, and so forth.

The main values to control are therefore:

- • Supersaturation value, as an index of the quantity of solute available for the growth of the crystal;

- • Total crystal surface in unit fluid mass, as an index of the capability of the solute to fix onto the crystal;

- • Retention time, as an index of the probability of a molecule of solute to come into contact with an existing crystal;

- • Flow pattern, again as an index of the probability of a molecule of solute to come into contact with an existing crystal (higher in laminar flow, lower in turbulent flow, but the reverse applies to the probability of contact).

The first value is a consequence of the physical characteristics of the solution, while the others define a difference between a well- and poorly designed crystallizer.

Crystal Size Distribution

The appearance and size range of a crystalline product is extremely important in crystallization. If further processing of the crystals is desired, large crystals with uniform size are important for washing, filtering, transportation, and storage. The importance lies in the fact that large crystals are easier to filter out of a solution than small crystals. Also, larger crystals have a smaller surface area to volume ratio, leading to a higher purity. This higher purity is due to less retention of mother liquor which contains impurities, and a smaller loss of yield when the crystals are washed to remove the mother liquor. The theoretical crystal size distribution can be estimated as a function of operating conditions with a fairly complicated mathematical process called population balance theory (using population balance equations).

Main Crystallization Processes

Some of the important factors influencing solubility are:

- Concentration

- Temperature

- Polarity

- Ionic Strength

So we may identify two main families of crystallization processes:

- Cooling crystallization

- Evaporative crystallization

This division is not really clear-cut, since hybrid systems exist, where cooling is performed through evaporation, thus obtaining at the same time a concentration of the solution.

A crystallization process often referred to in chemical engineering is the fractional crystallization. This is not a different process, rather a special application of one (or both) of the above.

Cooling Crystallization

Application

Most chemical compounds, dissolved in most solvents, show the so-called *direct* solubility that is, the solubility threshold increases with temperature.

So, whenever the conditions are favourable, crystal formation results from simply cooling the solution. Here *cooling* is a relative term: austenite crystals in a steel form well above 1000 °C. An example of this crystallization process is the production of Glauber's salt, a crystalline form of sodium sulfate. In the diagram, where equilibrium temperature is on the x-axis and equilibrium concentration (as mass percent of solute in saturated solution) in y-axis, it is clear that sulfate solubility quickly decreases below 32.5 °C. Assuming a saturated solution at 30 °C, by cooling it to 0 °C (note that this is possible thanks to the freezing-point depression), the precipitation of a mass of sulfate occurs corresponding to the change in solubility from 29% (equilibrium value at 30 °C) to approximately 4.5% (at 0 °C) – actually a larger crystal mass is precipitated, since sulfate entrains hydration water, and this has the side effect of increasing the final concentration.

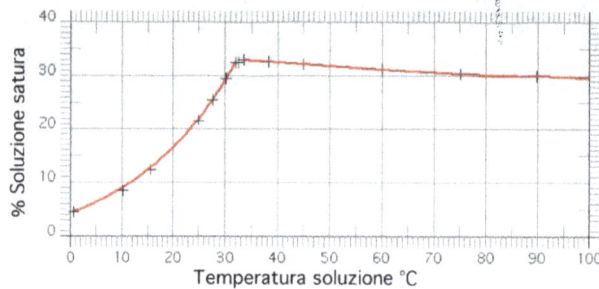

Solubility of the system $Na_2SO_4 - H_2O$

There are of course limitation in the use of cooling crystallization:

- Many solutes precipitate in hydrate form at low temperatures: in the previous example this is acceptable, and even useful, but it may be detrimental when, for example, the mass of water of hydration to reach a stable hydrate crystallization form is more than the available water: a single block of hydrate solute will be formed – this occurs in the case of calcium chloride);

- Maximum supersaturation will take place in the coldest points. These may be the heat exchanger tubes which are sensitive to scaling, and heat exchange may be greatly reduced or discontinued;

- A decrease in temperature usually implies an increase of the viscosity of a solution. Too high a viscosity may give hydraulic problems, and the laminar flow thus created may affect the crystallization dynamics.

- It is of course not applicable to compounds having *reverse* solubility, a term to indicate that solubility increases with temperature decrease (an example occurs with sodium sulfate where solubility is reversed above 32.5 °C).

Cooling Crystallizers

The simplest cooling crystallizers are tanks provided with a mixer for internal circulation, where temperature decrease is obtained by heat exchange with an intermediate fluid circulating in a jacket. These simple machines are used in batch processes, as in processing of pharmaceuticals and are prone to scaling. Batch processes normally provide a relatively variable quality of product along the batch.

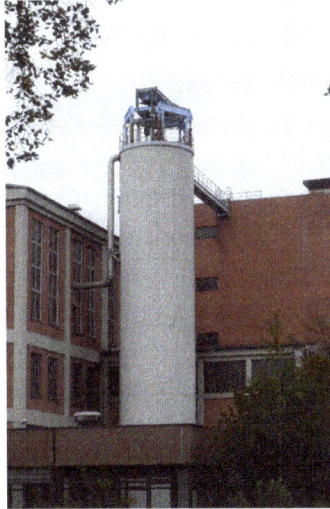

Vertical cooling crystallizer in a beet sugar factory

The *Swenson-Walker* crystallizer is a model, specifically conceived by Swenson Co. around 1920, having a semicylindric horizontal hollow trough in which a hollow screw conveyor or some hollow discs, in which a refrigerating fluid is circulated, plunge during rotation on a longitudinal axis. The refrigerating fluid is sometimes also circulated in a jacket around the trough. Crystals precipitate on the cold surfaces of the screw/discs, from which they are removed by scrapers and settle on the bottom of the trough. The screw, if provided, pushes the slurry towards a discharge port.

A common practice is to cool the solutions by flash evaporation: when a liquid at a given T_0 temperature is transferred in a chamber at a pressure P_1 such that the liquid saturation temperature T_1 at P_1 is lower than T_0, the liquid will release heat according to the temperature difference and a quantity of solvent, whose total latent heat of vaporization equals the difference in enthalpy. In simple words, the liquid is cooled by evaporating a part of it.

In the sugar industry vertical cooling crystallizers are used to exhaust the molasses in the last crystallization stage downstream of vacuum pans, prior to centrifugation. The massecuite enters the crystallizers at the top, and cooling water is pumped through pipes in counterflow.

Evaporative Crystallization

Another option is to obtain, at an approximately constant temperature, the precipitation of the crystals by increasing the solute concentration above the solubility threshold. To obtain this, the solute/solvent mass ratio is increased using the technique of evaporation. This process is of course insensitive to change in temperature (as long as hydration state remains unchanged).

All considerations on control of crystallization parameters are the same as for the cooling models.

Evaporative Crystallizers

Most industrial crystallizers are of the evaporative type, such as the very large sodium chloride and sucrose units, whose production accounts for more than 50% of the total world production of crystals. The most common type is the *forced circulation* (FC) model. A pumping device (a pump or an axial flow mixer) keeps the crystal slurry in homogeneous suspension throughout the tank, including the exchange surfaces; by controlling pump flow, control of the contact time of the crystal mass with the supersaturated solution is achieved, together with reasonable velocities at the exchange surfaces. The Oslo, mentioned above, is a refining of the evaporative forced circulation crystallizer, now equipped with a large crystals settling zone to increase the retention time (usually low in the FC) and to roughly separate heavy slurry zones from clear liquid.

DTB Crystallizer

DTB Crystallizer

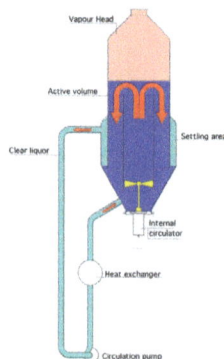

Schematic of DTB

Whichever the form of the crystallizer, to achieve an effective process control it is important to control the retention time and the crystal mass, to obtain the optimum conditions in terms of crystal specific surface and the fastest possible growth. This is achieved by a separation – to put it simply – of the crystals from the liquid mass, in order to manage the two flows in a different way. The practical way is to perform a gravity settling to be able to extract (and possibly recycle separately) the (almost) clear liquid, while managing the mass flow around the crystallizer to obtain a precise slurry density elsewhere. A typical example is the DTB (*Draft Tube and Baffle*) crystallizer, an idea of Richard Chisum Bennett (a Swenson engineer and later President of Swenson) at the end of the 1950s. The DTB crystallizer has an internal circulator, typically an axial flow mixer – yellow – pushing upwards in a draft tube while outside the crystallizer there is a settling area in an annulus; in it the exhaust solution moves upwards at a very low velocity, so that large crystals settle – and return to the main circulation – while only the fines, below a given grain size are extracted and eventually destroyed by increasing or decreasing temperature, thus creating additional supersaturation. A quasi-perfect control of all parameters is achieved. This crystallizer, and the derivative models (Krystal, CSC, etc.) could be the ultimate solution if not for a major limitation in the evaporative capacity, due to the limited diameter of the vapour head and the relatively low external circulation not allowing large amounts of energy to be supplied to the system.

Distillation

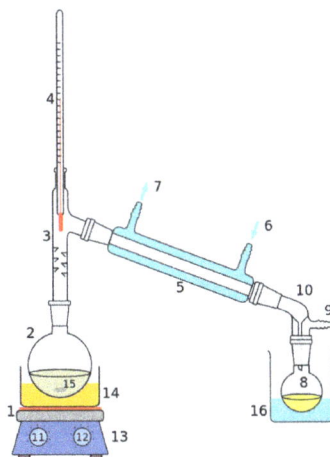

Laboratory display of distillation: **1:** A source of heat **2:** Still pot **3:** Still head **4:** Thermometer/Boiling point temperature **5:** Condenser **6:** Cooling water in **7:** Cooling water out **8:** Distillate/receiving flask **9:** Vacuum/gas inlet **10:** Still receiver **11:** Heat control **12:** Stirrer speed control **13:** Stirrer/heat plate **14:** Heating (Oil/sand) bath **15:** Stirring means e.g. (shown), boiling chips or mechanical stirrer **16:** Cooling bath.

Distillation is a process of separating the component or substances from a liquid mixture by selective evaporation and condensation. Distillation may result in essentially complete separation (nearly pure components), or it may be a partial separation that increases the concentration of selected components of the mixture. In either case the process exploits differences in the volatility of the mixture's components. In industrial chemistry, distillation is a unit operation of practically universal importance, but it is a physical separation process and not a chemical reaction.

Commercially, distillation has many applications. For example:

- In the fossil fuel industry distillation is a major class of operation in obtaining materials from crude oil for fuels and for chemical feedstocks.

- Distillation permits separation of air into its components — notably oxygen, nitrogen, and argon — for industrial use.

- In the field of industrial chemistry, large ranges of crude liquid products of chemical synthesis are distilled to separate them, either from other products, or from impurities, or from unreacted starting materials.

- Distillation of fermented products produces distilled beverages with a high alcohol content, or separates out other fermentation products of commercial value.

An installation for distillation, especially of alcohol, is a distillery. The distillation equipment is a still.

Distillation is a very old method of artificial desalination.

History

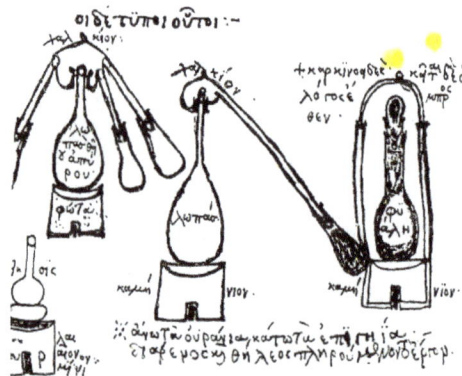

Distillation equipment used by the 3rd century Greek alchemist Zosimos of Panopolis, from the Byzantine Greek manuscript *Parisinus graces*.

Aristotle wrote about the process in his *Meteorologica* and even that "ordinary wine possesses a kind of exhalation, and that is why it gives out a flame". Later evidence of distillation comes from Greek alchemists working in Alexandria in the 1st century AD. Distilled water has been known since at least c. 200, when Alexander of Aphrodisias described the process. Distillation in China could have begun during the Eastern Han

Dynasty (1st–2nd centuries), but archaeological evidence indicates that actual distillation of beverages began in the Jin and Southern Song dynasties. A still was found in an archaeological site in Qinglong, Hebei province dating to the 12th century. Distilled beverages were more common during the Yuan dynasty. Arabs learned the process from the Alexandrians and used it extensively in their chemical experiments.

Clear evidence of the distillation of alcohol comes from the School of Salerno in the 12th century. Fractional distillation was developed by Tadeo Alderotti in the 13th century.

In 1500, German alchemist Hieronymus Braunschweig published *Liber de arte destillandi* (The Book of the Art of Distillation) the first book solely dedicated to the subject of distillation, followed in 1512 by a much expanded version. In 1651, John French published The Art of Distillation the first major English compendium of practice, though it has been claimed that much of it derives from Braunschweig's work. This includes diagrams with people in them showing the industrial rather than bench scale of the operation.

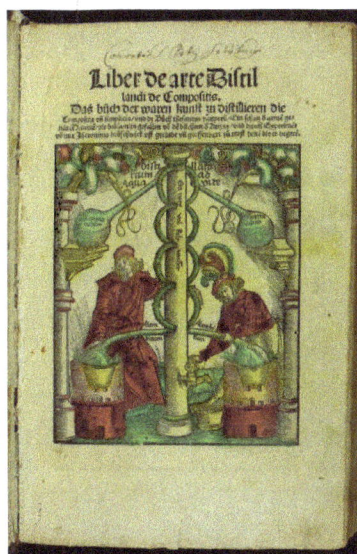

Hieronymus Brunschwig's *Liber de arte Distillandi de Compositis* (Strassburg, 1512) Chemical Heritage Foundation

A retort

Distillation

Old Ukrainian vodka still

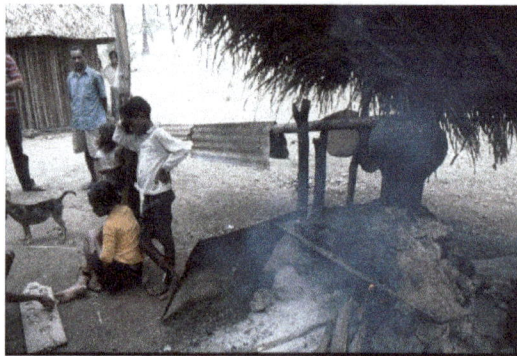
Simple liqueur distillation in East Timor

As alchemy evolved into the science of chemistry, vessels called retorts became used for distillations. Both alembics and retorts are forms of glassware with long necks pointing to the side at a downward angle which acted as air-cooled condensers to condense the distillate and let it drip downward for collection. Later, copper alembics were invented. Riveted joints were often kept tight by using various mixtures, for instance a dough made of rye flour. These alembics often featured a cooling system around the beak, using cold water for instance, which made the condensation of alcohol more efficient. These were called pot stills. Today, the retorts and pot stills have been largely supplanted by more efficient distillation methods in most industrial processes. However, the pot still is still widely used for the elaboration of some fine alcohols such as cognac, Scotch whisky, tequila and some vodkas. Pot stills made of various materials (wood, clay, stainless steel) are also used by bootleggers in various countries. Small pot stills are also sold for the domestic production of flower water or essential oils.

Early forms of distillation were batch processes using one vaporization and one condensation. Purity was improved by further distillation of the condensate. Greater volumes were processed by simply repeating the distillation. Chemists were reported to carry out as many as 500 to 600 distillations in order to obtain a pure compound.

In the early 19th century the basics of modern techniques including pre-heating and reflux were developed, particularly by the French, then in 1830 a British Patent was issued to Aeneas Coffey for a whisky distillation column, which worked continuously and may be regarded as the archetype of modern petrochemical units. In 1877, Ernest Solvay was granted a U.S. Patent for a tray column for ammonia distillation and the same and subsequent years saw developments of this theme for oil and spirits.

With the emergence of chemical engineering as a discipline at the end of the 19th century, scientific rather than empirical methods could be applied. The developing petroleum industry in the early 20th century provided the impetus for the development of accurate design methods such as the McCabe–Thiele method and the Fenske equation. The availability of powerful computers has also allowed direct computer simulation of distillation columns.

Applications of Distillation

The application of distillation can roughly be divided in four groups: laboratory scale, industrial distillation, distillation of herbs for perfumery and medicinals (herbal distillate), and food processing. The latter two are distinctively different from the former two in that in the processing of beverages and herbs, the distillation is not used as a true purification method but more to transfer all volatiles from the source materials to the distillate.

The main difference between laboratory scale distillation and industrial distillation is that laboratory scale distillation is often performed batch-wise, whereas industrial distillation often occurs continuously. In batch distillation, the composition of the source material, the vapors of the distilling compounds and the distillate change during the distillation. In batch distillation, a still is charged (supplied) with a batch of feed mixture, which is then separated into its component fractions which are collected sequentially from most volatile to less volatile, with the bottoms (remaining least or non-volatile fraction) removed at the end. The still can then be recharged and the process repeated.

In continuous distillation, the source materials, vapors, and distillate are kept at a constant composition by carefully replenishing the source material and removing fractions from both vapor and liquid in the system. This results in a better control of the separation process.

Idealized Distillation Model

The boiling point of a liquid is the temperature at which the vapor pressure of the liquid equals the pressure around the liquid, enabling bubbles to form without being crushed.

A special case is the normal boiling point, where the vapor pressure of the liquid equals the ambient atmospheric pressure.

It is a common misconception that in a liquid mixture at a given pressure, each component boils at the boiling point corresponding to the given pressure and the vapors of each component will collect separately and purely. This, however, does not occur even in an idealized system. Idealized models of distillation are essentially governed by Raoult's law and Dalton's law, and assume that vapor–liquid equilibria are attained.

Raoult's law states that the vapor pressure of a solution is dependent on 1) the vapor pressure of each chemical component in the solution and 2) the fraction of solution each component makes up a.k.a. the mole fraction. This law applies to ideal solutions, or solutions that have different components but whose molecular interactions are the same as or very similar to pure solutions.

Dalton's law states that the total pressure is the sum of the partial pressures of each individual component in the mixture. When a multi-component liquid is heated, the vapor pressure of each component will rise, thus causing the total vapor pressure to rise. When the total vapor pressure reaches the pressure surrounding the liquid, boiling occurs and liquid turns to gas throughout the bulk of the liquid. Note that a mixture with a given composition has one boiling point at a given pressure, when the components are mutually soluble. A mixture of constant composition does not have multiple boiling points.

An implication of one boiling point is that lighter components never cleanly "boil first". At boiling point, all volatile components boil, but for a component, its percentage in the vapor is the same as its percentage of the total vapor pressure. Lighter components have a higher partial pressure and thus are concentrated in the vapor, but heavier volatile components also have a (smaller) partial pressure and necessarily evaporate also, albeit being less concentrated in the vapor. Indeed, batch distillation and fractionation succeed by varying the composition of the mixture. In batch distillation, the batch evaporates, which changes its composition; in fractionation, liquid higher in the fractionation column contains more lights and boils at lower temperatures. Therefore, starting from a given mixture, it appears to have a boiling range instead of a boiling *point*, although this is because its composition changes: each intermediate mixture has its own, singular boiling point.

The idealized model is accurate in the case of chemically similar liquids, such as benzene and toluene. In other cases, severe deviations from Raoult's law and Dalton's law are observed, most famously in the mixture of ethanol and water. These compounds, when heated together, form an azeotrope, which is a composition with a boiling point higher or lower than the boiling point of each separate liquid. Virtually all liquids, when mixed and heated, will display azeotropic behaviour. Although there are computational methods that can be used to estimate the behavior of a mixture of arbitrary components, the only way to obtain accurate vapor–liquid equilibrium data is by measurement.

It is not possible to *completely* purify a mixture of components by distillation, as this would require each component in the mixture to have a zero partial pressure. If ultra-pure products are the goal, then further chemical separation must be applied. When a binary mixture is evaporated and the other component, e.g. a salt, has zero partial pressure for practical purposes, the process is simpler and is called evaporation in engineering.

Batch Distillation

Heating an ideal mixture of two volatile substances A and B (with A having the higher volatility, or lower boiling point) in a batch distillation setup (such as in an apparatus depicted in the opening figure) until the mixture is boiling results in a vapor above the liquid which contains a mixture of A and B. The ratio between A and B in the vapor will be different from the ratio in the liquid: the ratio in the liquid will be determined by how the original mixture was prepared, while the ratio in the vapor will be enriched in the more volatile compound, A. The vapor goes through the condenser and is removed from the system. This in turn means that the ratio of compounds in the remaining liquid is now different from the initial ratio (i.e., more enriched in B than the starting liquid).

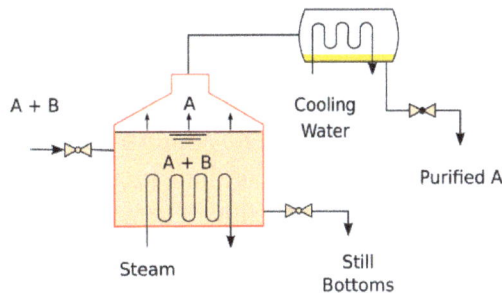

A batch still showing the separation of A and B.

The result is that the ratio in the liquid mixture is changing, becoming richer in component B. This causes the boiling point of the mixture to rise, which in turn results in a rise in the temperature in the vapor, which results in a changing ratio of A : B in the gas phase (as distillation continues, there is an increasing proportion of B in the gas phase). This results in a slowly changing ratio A : B in the distillate.

If the difference in vapor pressure between the two components A and B is large (generally expressed as the difference in boiling points), the mixture in the beginning of the distillation is highly enriched in component A, and when component A has distilled off, the boiling liquid is enriched in component B.

Continuous Distillation

Continuous distillation is an ongoing distillation in which a liquid mixture is continuously (without interruption) fed into the process and separated fractions are removed continuously as output streams occur over time during the operation. Continuous distillation produces a minimum of two output fractions, including at least one volatile distillate fraction, which has boiled and been separately captured as a vapor, and then condensed to a liquid. There is always a bottoms (or residue) fraction, which is the least volatile residue that has not been separately captured as a condensed vapor.

Continuous distillation differs from batch distillation in the respect that concentrations should not change over time. Continuous distillation can be run at a steady state for an arbitrary amount of time. For any source material of specific composition, the main variables that affect the purity of products in continuous distillation are the reflux ratio and the number of theoretical equilibrium stages, in practice determined by the number of trays or the height of packing. Reflux is a flow from the condenser back to the column, which generates a recycle that allows a better separation with a given number of trays. Equilibrium stages are ideal steps where compositions achieve vapor–liquid equilibrium, repeating the separation process and allowing better separation given a reflux ratio. A column with a high reflux ratio may have fewer stages, but it refluxes a large amount of liquid, giving a wide column with a large holdup. Conversely, a column with a low reflux ratio must have a large number of stages, thus requiring a taller column.

General Improvements

Both batch and continuous distillations can be improved by making use of a fractionating column on top of the distillation flask. The column improves separation by providing a larger surface area for the vapor and condensate to come into contact. This helps it remain at equilibrium for as long as possible. The column can even consist of small subsystems ('trays' or 'dishes') which all contain an enriched, boiling liquid mixture, all with their own vapor–liquid equilibrium.

There are differences between laboratory-scale and industrial-scale fractionating columns, but the principles are the same. Examples of laboratory-scale fractionating columns (in increasing efficiency) include

- Air condenser
- Vigreux column (usually laboratory scale only)
- Packed column (packed with glass beads, metal pieces, or other chemically inert material)
- Spinning band distillation system.

Laboratory Scale Distillation

Laboratory scale distillations are almost exclusively run as batch distillations. The device used in distillation, sometimes referred to as a *still*, consists at a minimum of a reboiler or *pot* in which the source material is heated, a condenser in which the heated vapour is cooled back to the liquid state, and a receiver in which the concentrated or purified liquid, called the distillate, is collected. Several laboratory scale techniques for distillation exist.

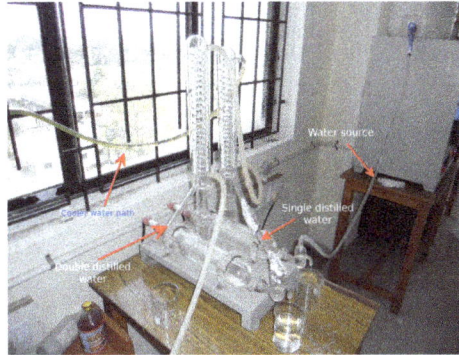

Typical laboratory distillation unit

Simple Distillation

In simple distillation, the vapor is immediately channeled into a condenser. Consequently, the distillate is not pure but rather its composition is identical to the composition of the vapors at the given temperature and pressure. That concentration follows Raoult's law.

As a result, simple distillation is effective only when the liquid boiling points differ greatly (rule of thumb is 25 °C) or when separating liquids from non-volatile solids or oils. For these cases, the vapor pressures of the components are usually different enough that the distillate may be sufficiently pure for its intended purpose.

Fractional Distillation

For many cases, the boiling points of the components in the mixture will be sufficiently close that Raoult's law must be taken into consideration. Therefore, fractional distillation must be used in order to separate the components by repeated vaporization-condensation cycles within a packed fractionating column. This separation, by successive distillations, is also referred to as rectification.

As the solution to be purified is heated, its vapors rise to the fractionating column. As it rises, it cools, condensing on the condenser walls and the surfaces of the packing material. Here, the condensate continues to be heated by the rising hot vapors; it vaporizes once more. However, the composition of the fresh vapors are determined once again by Raoult's law. Each vaporization-condensation cycle (called a *theoretical plate*) will yield a purer solution of the more volatile component. In reality, each cycle at a given temperature does not occur at exactly the same position in the fractionating column; *theoretical plate* is thus a concept rather than an accurate description.

More theoretical plates lead to better separations. A spinning band distillation system uses a spinning band of Teflon or metal to force the rising vapors into close contact with the descending condensate, increasing the number of theoretical plates.

Steam Distillation

Like vacuum distillation, steam distillation is a method for distilling compounds which are heat-sensitive. The temperature of the steam is easier to control than the surface of a heating element, and allows a high rate of heat transfer without heating at a very high temperature. This process involves bubbling steam through a heated mixture of the raw material. By Raoult's law, some of the target compound will vaporize (in accordance with its partial pressure). The vapor mixture is cooled and condensed, usually yielding a layer of oil and a layer of water.

Steam distillation of various aromatic herbs and flowers can result in two products; an essential oil as well as a watery herbal distillate. The essential oils are often used in perfumery and aromatherapy while the watery distillates have many applications in aromatherapy, food processing and skin care.

Dimethyl sulfoxide usually boils at 189 °C. Under a vacuum, it distills off into the receiver at only 70 °C.

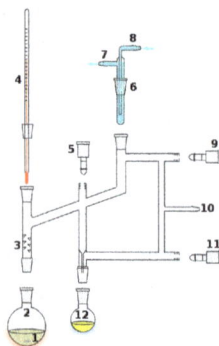

Perkin triangle distillation setup
1: Stirrer bar/anti-bumping granules **2:** Still pot **3:** Fractionating column **4:** Thermometer/Boiling point temperature **5:** Teflon tap 1 **6:** Cold finger **7:** Cooling water out **8:** Cooling water in **9:** Teflon tap 2 **10:** Vacuum/gas inlet **11:** Teflon tap 3 **12:** Still receiver

Vacuum Distillation

Some compounds have very high boiling points. To boil such compounds, it is often better to lower the pressure at which such compounds are boiled instead of increasing the temperature. Once the pressure is lowered to the vapor pressure of the compound (at the given temperature), boiling and the rest of the distillation process can commence. This technique is referred to as vacuum distillation and it is commonly found in the laboratory in the form of the rotary evaporator.

This technique is also very useful for compounds which boil beyond their decomposition temperature at atmospheric pressure and which would therefore be decomposed by any attempt to boil them under atmospheric pressure.

Molecular distillation is vacuum distillation below the pressure of 0.01 torr. 0.01 torr is one order of magnitude above high vacuum, where fluids are in the free molecular flow regime, i.e. the mean free path of molecules is comparable to the size of the equipment. The gaseous phase no longer exerts significant pressure on the substance to be evaporated, and consequently, rate of evaporation no longer depends on pressure. That is, because the continuum assumptions of fluid dynamics no longer apply, mass transport is governed by molecular dynamics rather than fluid dynamics. Thus, a short path between the hot surface and the cold surface is necessary, typically by suspending a hot plate covered with a film of feed next to a cold plate with a line of sight in between. Molecular distillation is used industrially for purification of oils.

Air-sensitive Vacuum Distillation

Some compounds have high boiling points as well as being air sensitive. A simple vacuum distillation system as exemplified above can be used, whereby the vacuum is replaced with an inert gas after the distillation is complete. However, this is a less satisfactory system if one desires to collect fractions under a reduced pressure. To do this a "cow" or "pig" adaptor can be added to the end of the condenser, or for better results or for very air sensitive compounds a Perkin triangle apparatus can be used.

The Perkin triangle, has means via a series of glass or Teflon taps to allows fractions to be isolated from the rest of the still, without the main body of the distillation being removed from either the vacuum or heat source, and thus can remain in a state of reflux. To do this, the sample is first isolated from the vacuum by means of the taps, the vacuum over the sample is then replaced with an inert gas (such as nitrogen or argon) and can then be stoppered and removed. A fresh collection vessel can then be added to the system, evacuated and linked back into the distillation system via the taps to collect a second fraction, and so on, until all fractions have been collected.

4

Short Path Distillation

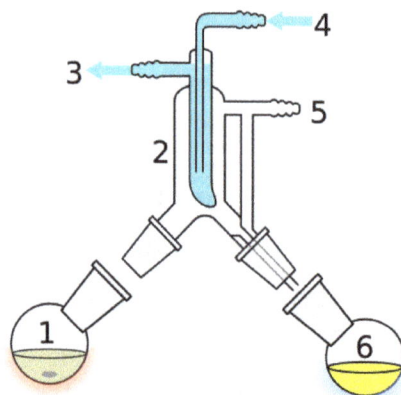

Short path vacuum distillation apparatus with vertical condenser (cold finger), to minimize the distillation path; **1:** Still pot with stirrer bar/anti-bumping granules **2:** Cold finger – bent to direct condensate **3:** Cooling water out **4:** cooling water in **5:** Vacuum/gas inlet **6:** Distillate flask/distillate.

Short path distillation is a distillation technique that involves the distillate travelling a short distance, often only a few centimeters, and is normally done at reduced pressure. A classic example would be a distillation involving the distillate travelling from one glass bulb to another, without the need for a condenser separating the two chambers. This technique is often used for compounds which are unstable at high temperatures or to purify small amounts of compound. The advantage is that the heating temperature can be considerably lower (at reduced pressure) than the boiling point of the liquid at standard pressure, and the distillate only has to travel a short distance before condensing. A short path ensures that little compound is lost on the sides of the apparatus. The Kugelrohr is a kind of a short path distillation apparatus which often contain multiple chambers to collect distillate fractions.

Zone Distillation

Zone distillation is a distillation process in long container with partial melting of refined matter in moving liquid zone and condensation of vapor in the solid phase at condensate pulling in cold area. The process is worked in theory. When zone heater is moving from the top to the bottom of the container then solid condensate with irregular impurity distribution is forming. Then most pure part of the condensate may be extracted as product. The process may be iterated many times by moving (without turnover) the received condensate to the bottom part of the container on the place of refined matter. The irregular impurity distribution in the condensate (that is efficiency of purification) increases with number of repetitions of the process. Zone distillation is a distillation analog of zone recrystallization. Impurity distribution in the condensate is described by known equations of zone recrystallization with various numbers of iteration of process – with replacement distribution efficient k of crystallization on separation factor α of distillation.

Other Types

- The process of reactive distillation involves using the reaction vessel as the still. In this process, the product is usually significantly lower-boiling than its reactants. As the product is formed from the reactants, it is vaporized and removed from the reaction mixture. This technique is an example of a continuous vs. a batch process; advantages include less downtime to charge the reaction vessel with starting material, and less workup. Distillation "over a reactant" could be classified as a reactive distillation. It is typically used to remove volatile impurity from the distallation feed. For example, a little lime may be added to remove carbon dioxide from water followed by a second distillation with a little sulphuric acid added to remove traces of ammonia.

- Catalytic distillation is the process by which the reactants are catalyzed while being distilled to continuously separate the products from the reactants. This method is used to assist equilibrium reactions reach completion.

- Pervaporation is a method for the separation of mixtures of liquids by partial vaporization through a non-porous membrane.

- Extractive distillation is defined as distillation in the presence of a miscible, high boiling, relatively non-volatile component, the solvent, that forms no azeotrope with the other components in the mixture.

- Flash evaporation (or partial evaporation) is the partial vaporization that occurs when a saturated liquid stream undergoes a reduction in pressure by passing through a throttling valve or other throttling device. This process is one of the simplest unit operations, being equivalent to a distillation with only one equilibrium stage.

- Codistillation is distillation which is performed on mixtures in which the two compounds are not miscible.

The unit process of evaporation may also be called "distillation":

- In rotary evaporation a vacuum distillation apparatus is used to remove bulk solvents from a sample. Typically the vacuum is generated by a water aspirator or a membrane pump.

- In a kugelrohr a short path distillation apparatus is typically used (generally in combination with a (high) vacuum) to distill high boiling (> 300 °C) compounds. The apparatus consists of an oven in which the compound to be distilled is placed, a receiving portion which is outside of the oven, and a means of rotating the sample. The vacuum is normally generated by using a high vacuum pump.

Other uses:

- Dry distillation or destructive distillation, despite the name, is not truly distillation, but rather a chemical reaction known as pyrolysis in which solid substances are heated in an inert or reducing atmosphere and any volatile fractions, containing high-boiling liquids and products of pyrolysis, are collected. The destructive distillation of wood to give methanol is the root of its common name – *wood alcohol*.

- Freeze distillation is an analogous method of purification using freezing instead of evaporation. It is not truly distillation, but a recrystallization where the product is the mother liquor, and does not produce products equivalent to distillation. This process is used in the production of ice beer and ice wine to increase ethanol and sugar content, respectively. It is also used to produce applejack. Unlike distillation, freeze distillation concentrates poisonous congeners rather than removing them; As a result, many countries prohibit such applejack as a health measure. However, reducing methanol with the absorption of 4A molecular sieve is a practical method for production. Also, distillation by evaporation can separate these since they have different boiling points.

Azeotropic Distillation

Interactions between the components of the solution create properties unique to the solution, as most processes entail nonideal mixtures, where Raoult's law does not hold. Such interactions can result in a constant-boiling azeotrope which behaves as if it were a pure compound (i.e., boils at a single temperature instead of a range). At an azeotrope, the solution contains the given component in the same proportion as the vapor, so that evaporation does not change the purity, and distillation does not effect separation. For example, ethyl alcohol and water form an azeotrope of 95.6% at 78.1 °C.

If the azeotrope is not considered sufficiently pure for use, there exist some techniques to break the azeotrope to give a pure distillate. This set of techniques are known as azeotropic distillation. Some techniques achieve this by "jumping" over the azeotropic composition (by adding another component to create a new azeotrope, or by varying the pressure). Others work by chemically or physically removing or sequestering the impurity. For example, to purify ethanol beyond 95%, a drying agent (or desiccant, such as potassium carbonate) can be added to convert the soluble water into insoluble water of crystallization. Molecular sieves are often used for this purpose as well.

Immiscible liquids, such as water and toluene, easily form azeotropes. Commonly, these azeotropes are referred to as a low boiling azeotrope because the boiling point of the azeotrope is lower than the boiling point of either pure component. The temperature and composition of the azeotrope is easily predicted from the vapor pressure of the pure components, without use of Raoult's law. The azeotrope is easily broken in a distillation set-up by using a liquid–liquid separator (a decanter) to separate the two liquid layers that are condensed overhead. Only one of the two liquid layers is refluxed to the distillation set-up.

High boiling azeotropes, such as a 20 weight percent mixture of hydrochloric acid in water, also exist. As implied by the name, the boiling point of the azeotrope is greater than the boiling point of either pure component.

To break azeotropic distillations and cross distillation boundaries, such as in the DeRosier Problem, it is necessary to increase the composition of the light key in the distillate.

Breaking an Azeotrope with Unidirectional Pressure Manipulation

The boiling points of components in an azeotrope overlap to form a band. By exposing an azeotrope to a vacuum or positive pressure, it's possible to bias the boiling point of one component away from the other by exploiting the differing vapour pressure curves of each; the curves may overlap at the azeotropic point, but are unlikely to be remain identical further along the pressure axis either side of the azeotropic point. When the bias is great enough, the two boiling points no longer overlap and so the azeotropic band disappears.

This method can remove the need to add other chemicals to a distillation, but it has two potential drawbacks.

Under negative pressure, power for a vacuum source is needed and the reduced boiling points of the distillates requires that the condenser be run cooler to prevent distillate vapours being lost to the vacuum source. Increased cooling demands will often require additional energy and possibly new equipment or a change of coolant.

Alternatively, if positive pressures are required, standard glassware can not be used, energy must be used for pressurization and there is a higher chance of side reactions occurring in the distillation, such as decomposition, due to the higher temperatures required to effect boiling.

A unidirectional distillation will rely on a pressure change in one direction, either positive or negative.

Pressure-swing Distillation

Pressure-swing distillation is essentially the same as the unidirectional distillation used to break azeotropic mixtures, but here both positive and negative pressures may be employed.

This improves the selectivity of the distillation and allows a chemist to optimize distillation by avoiding extremes of pressure and temperature that waste energy. This is particularly important in commercial applications.

One example of the application of pressure-swing distillation is during the industrial purification of ethyl acetate after its catalytic synthesis from ethanol.

Industrial Distillation

Large scale industrial distillation applications include both batch and continuous fractional, vacuum, azeotropic, extractive, and steam distillation. The most widely used industrial applications of continuous, steady-state fractional distillation are in petroleum refineries, petrochemical and chemical plants and natural gas processing plants.

Typical industrial distillation towers

To control and optimize such industrial distillation, a standardized laboratory method, ASTM D86, is established. This test method extends to the atmospheric distillation of petroleum products using a laboratory batch distillation unit to quantitatively determine the boiling range characteristics of petroleum products.

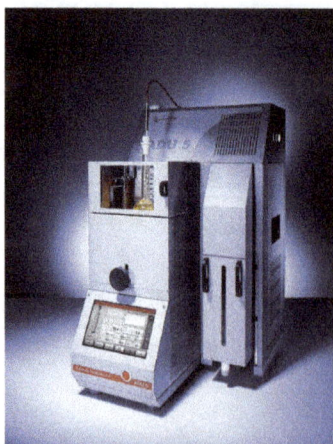

Automatic Distillation Unit for the determination of the boiling range of petroleum products at atmospheric pressure

Industrial distillation is typically performed in large, vertical cylindrical columns known as distillation towers or distillation columns with diameters ranging from about 65 centimeters to 16 meters and heights ranging from about 6 meters to 90 meters or more. When the process feed has a diverse composition, as in distilling crude oil, liquid outlets at intervals up the column allow for the withdrawal of different *fractions* or products having different boiling points or boiling ranges. The "lightest" products (those with the lowest boiling point) exit from the top of the columns and the "heaviest" products (those with the highest boiling point) exit from the bottom of the column and are often called the bottoms.

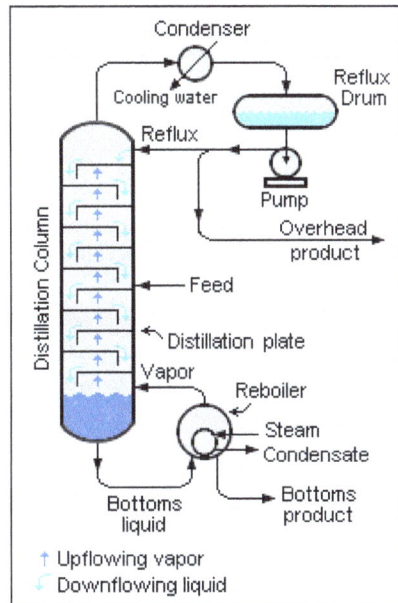

Diagram of a typical industrial distillation tower

Industrial towers use reflux to achieve a more complete separation of products. Reflux refers to the portion of the condensed overhead liquid product from a distillation or fractionation tower that is returned to the upper part of the tower as shown in the schematic diagram of a typical, large-scale industrial distillation tower. Inside the tower, the downflowing reflux liquid provides cooling and condensation of the upflowing vapors thereby increasing the efficiency of the distillation tower. The more reflux that is provided for a given number of theoretical plates, the better the tower's separation of lower boiling materials from higher boiling materials. Alternatively, the more reflux that is provided for a given desired separation, the fewer the number of theoretical plates required. Chemical engineers must choose what combination of reflux rate and number of plates is both economically and physically feasible for the products purified in the distillation column.

Such industrial fractionating towers are also used in cryogenic air separation, producing liquid oxygen, liquid nitrogen, and high purity argon. Distillation of chlorosilanes also enables the production of high-purity silicon for use as a semiconductor.

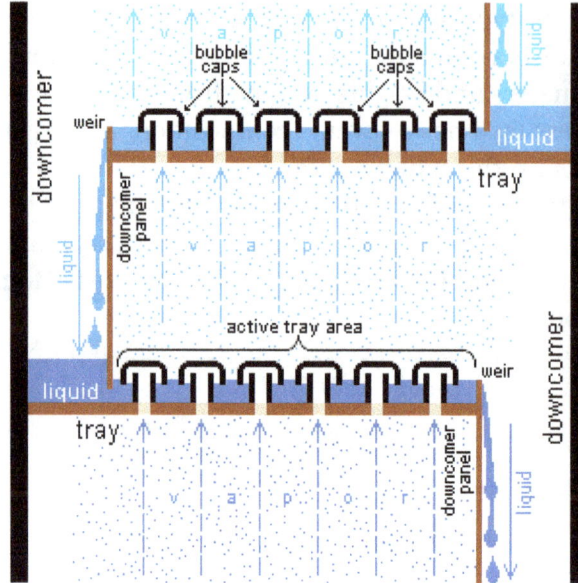

Section of an industrial distillation tower showing detail of trays with bubble caps

Design and operation of a distillation tower depends on the feed and desired products. Given a simple, binary component feed, analytical methods such as the McCabe–Thiele method or the Fenske equation can be used. For a multi-component feed, simulation models are used both for design and operation. Moreover, the efficiencies of the vapor–liquid contact devices (referred to as "plates" or "trays") used in distillation towers are typically lower than that of a theoretical 100% efficient equilibrium stage. Hence, a distillation tower needs more trays than the number of theoretical vapor–liquid equilibrium stages. A variety of models have been postulated to estimate tray efficiencies.

In modern industrial uses, a packing material is used in the column instead of trays when low pressure drops across the column are required. Other factors that favor packing are: vacuum systems, smaller diameter columns, corrosive systems, systems prone to foaming, systems requiring low liquid holdup, and batch distillation. Conversely, factors that favor plate columns are: presence of solids in feed, high liquid rates, large column diameters, complex columns, columns with wide feed composition variation, columns with a chemical reaction, absorption columns, columns limited by foundation weight tolerance, low liquid rate, large turn-down ratio and those processes subject to process surges.

This packing material can either be random dumped packing (1–3" wide) such as Raschig rings or structured sheet metal. Liquids tend to wet the surface of the packing and the vapors pass across this wetted surface, where mass transfer takes place. Unlike conventional tray distillation in which every tray represents a separate point of vapor–liquid equilibrium, the vapor–liquid equilibrium curve in a packed column is continuous. However, when modeling packed columns, it is useful to compute a number of "theoretical stages" to denote the separation efficiency of the packed column with respect to more traditional trays. Differently shaped packings have different surface areas and void space between packings. Both of these factors affect packing performance.

Large-scale, industrial vacuum distillation column

Another factor in addition to the packing shape and surface area that affects the performance of random or structured packing is the liquid and vapor distribution entering the packed bed. The number of theoretical stages required to make a given separation is calculated using a specific vapor to liquid ratio. If the liquid and vapor are not evenly distributed across the superficial tower area as it enters the packed bed, the liquid to vapor ratio will not be correct in the packed bed and the required separation will not be achieved. The packing will appear to not be working properly. The height equivalent to a theoretical plate (HETP) will be greater than expected. The problem is not the packing itself but the mal-distribution of the fluids entering the packed bed. Liquid mal-distribution is more frequently the problem than vapor. The design of the liquid distributors used to introduce the feed and reflux to a packed bed is critical to making the packing perform to it maximum efficiency. Methods of evaluating the effectiveness of a liquid distributor to evenly distribute the liquid entering a packed bed can be found in references. Considerable work as been done on this topic by Fractionation Research, Inc. (commonly known as FRI).

Multi-effect Distillation

The goal of multi-effect distillation is to increase the energy efficiency of the process, for use in desalination, or in some cases one stage in the production of ultrapure water. The number of effects is inversely proportional to the kW·h/m³ of water recovered figure, and refers to the volume of water recovered per unit of energy compared with single-effect distillation. One effect is roughly 636 kW·h/m³.

- Multi-stage flash distillation Can achieve more than 20 effects with thermal energy input, as mentioned in the article.

- Vapor compression evaporation Commercial large-scale units can achieve around 72 effects with electrical energy input, according to manufacturers.

There are many other types of multi-effect distillation processes, including one referred to as simply multi-effect distillation (MED), in which multiple chambers, with intervening heat exchangers, are employed.

Distillation in Food Processing

Distilled Beverages

Carbohydrate-containing plant materials are allowed to ferment, producing a dilute solution of ethanol in the process. Spirits such as whiskey and rum are prepared by distilling these dilute solutions of ethanol. Components other than ethanol, including water, esters, and other alcohols, are collected in the condensate, which account for the flavor of the beverage. Some of these beverages are then stored in barrels or other containers to acquire more flavor compounds and characteristic flavors.

Gallery

	Chemistry in its beginnings used retorts as laboratory equipment exclusively for distillation processes.
	A simple set-up to distill dry and oxygen-free toluene.
	Diagram of an industrial-scale vacuum distillation column as commonly used in oil refineries

Membrane Technology

Membrane technology covers all engineering approaches for the transport of substances between two fractions with the help of permeable membranes. In general, mechanical separation processes for separating gaseous or liquid streams use membrane technology.

Applications

Ultrafiltration for a swimming pool

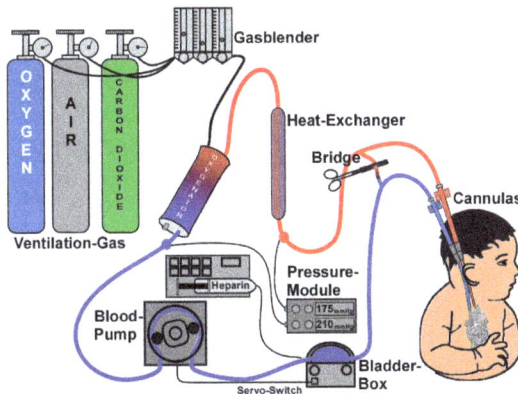

Venous-arterial extracorporeal membrane oxygenation scheme

Membrane separation processes operate without heating and therefore use less energy than conventional thermal separation processes such as distillation, sublimation or crystallization. The separation process is purely physical and both fractions (permeate and retentate) can be used. Cold separation using membrane technology is widely used in the food technology, biotechnology and pharmaceutical industries. Furthermore, using membranes enables separations to take place that would be impossible using thermal separation methods. For example, it is impossible to separate the constituents of azeotropic liquids or solutes which form isomorphic crystals by distillation or recrystallization but such separations can be achieved using membrane technology. Depending on the type of membrane, the selective separation of certain individual substances or substance mixtures is possible. Important technical applications include the production of drinking water by reverse osmosis (worldwide approximately 7 million cubic metres annually), filtrations in the food industry, the recovery of organic vapours such as petro-chemical vapour recovery and the electrolysis for chlorine production.

In waste water treatment, membrane technology is becoming increasingly important. With the help of ultra/microfiltration it is possible to remove particles, colloids and macromolecules, so that waste-water can be disinfected in this way. This is needed if waste-water is discharged into sensitive waters especially those designated for contact water-sports and recreation.

About half of the market is in medical applications such as use in artificial kidneys to remove toxic substances by hemodialysis and as artificial lung for bubble-free supply of oxygen in the blood.

The importance of membrane technology is growing in the field of environmental protection (NanoMemPro IPPC Database). Even in modern energy recovery techniques membranes are increasingly used, for example in fuel cells and in osmotic power plants.

Mass Transfer

Two basic models can be distinguished for mass transfer through the membrane:

- the *solution-diffusion model* and
- the *hydrodynamic model*.

In real membranes, these two transport mechanisms certainly occur side by side, especially during ultra-filtration.

Solution-diffusion Model

In the solution-diffusion model, transport occurs only by diffusion. The component that needs to be transported must first be dissolved in the membrane. The general approach of the solution-diffusion model is to assume that the chemical potential of the feed and permeate fluids are in equilibrium with the adjacent membrane surfaces such that appropriate expressions for the chemical potential in the fluid and membrane phases can be equated at the solution-membrane interface. This principle is more important for *dense* membranes without natural pores such as those used for reverse osmosis and in fuel cells. During the filtration process a boundary layer forms on the membrane. This concentration gradient is created by molecules which cannot pass through the membrane. The effect is referred as concentration polarization and, occurring during the filtration, leads to a reduced trans-membrane flow (flux). Concentration polarization is, in principle, reversible by cleaning the membrane which results in the initial flux being almost totally restored. Using a tangential flow to the membrane (cross-flow filtration) can also minimize concentration polarization.

Hydrodynamic Model

Transport through pores – in the simplest case – is done convectively. This requires the size of the pores to be smaller than the diameter of the two separate components. Membranes which function according to this principle are used mainly in micro- and ultrafiltration. They are used to separate macromolecules from solutions, colloids from a dispersion or remove bacteria. During this process the retained particles or molecules form a pulpy mass (filter cake) on the membrane, and this blockage of the membrane hampers the filtration. This blockage can be reduced by the use of the cross-flow method (cross-flow filtration). Here, the liquid to be filtered flows along the front of the membrane and is separated by the pressure difference between the front and back of the membrane into retentate (the flowing concentrate) on the front and permeate (filtrate) on the back. The tangential flow on the front creates a shear stress that cracks the filter cake and reduces the fouling.

Membrane Operations

According to the driving force of the operation it is possible to distinguish:

- Pressure driven operations
    - microfiltration
    - ultrafiltration
    - nanofiltration
    - reverse osmosis
- Concentration driven operations
    - dialysis
    - pervaporation
    - forward osmosis
    - artificial lung
    - gas separation
- Operations in an electric potential gradient
    - electrodialysis
    - membrane electrolysis e.g. chloralkali process
    - electrodeionization
    - electrofiltration
    - fuel cell

- Operations in a temperature gradient

  o membrane distillation

Membrane Shapes and Flow Geometries

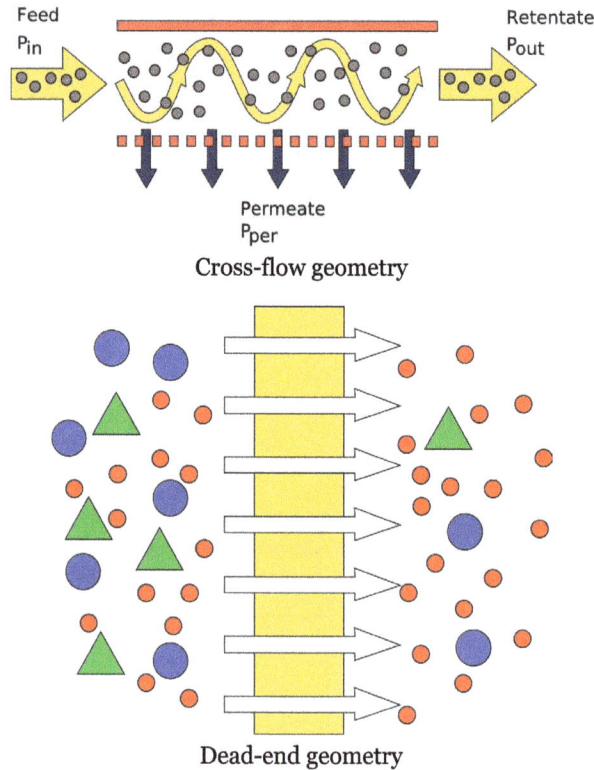

Cross-flow geometry

Dead-end geometry

There are two main flow configurations of membrane processes: cross-flow (or) tangential flow and dead-end filtrations. In cross-flow filtration the feed flow is tangential to the surface of membrane, retentate is removed from the same side further downstream, whereas the permeate flow is tracked on the other side. In dead-end filtration the direction of the fluid flow is normal to the membrane surface. Both flow geometries offer some advantages and disadvantages. Generally, dead-end filtration is used for feasibility studies on a laboratory scale. The dead-end membranes are relatively easy to fabricate which reduces the cost of the separation process. The dead-end membrane separation process is easy to implement and the process is usually cheaper than cross-flow membrane filtration. The dead-end filtration process is usually a batch-type process, where the filtering solution is loaded (or slowly fed) into the membrane device, which then allows passage of some particles subject to the driving force. The main disadvantage of a dead end filtration is the extensive membrane fouling and concentration polarization. The fouling is usually induced faster at higher driving forces. Membrane fouling and particle retention in a feed solution also builds up a concentration gradients and particle back flow (concentration polarization). The tangential flow devices are more cost and labor-intensive, but they are less susceptible to fouling due to the sweeping effects and high shear rates of the passing flow. The most commonly used synthetic membrane devices (modules) are flat sheets/plates, spiral wounds, and hollow fibers.

Flat plates are usually constructed as circular thin flat membrane surfaces to be used in dead-end geometry modules. Spiral wounds are constructed from similar flat membranes but in the form of a "pocket" containing two membrane sheets separated by a highly porous support plate. Several such pockets are then wound around a tube to create a tangential flow geometry and to reduce membrane fouling. hollow fiber modules consist of an assembly of self-supporting fibers with dense skin separation layers, and a more open matrix helping to withstand pressure gradients and maintain structural integrity. The hollow fiber modules can contain up to 10,000 fibers ranging from 200 to 2500 μm in diameter; The main advantage of hollow fiber modules is very large surface area within an enclosed volume, increasing the efficiency of the separation process.

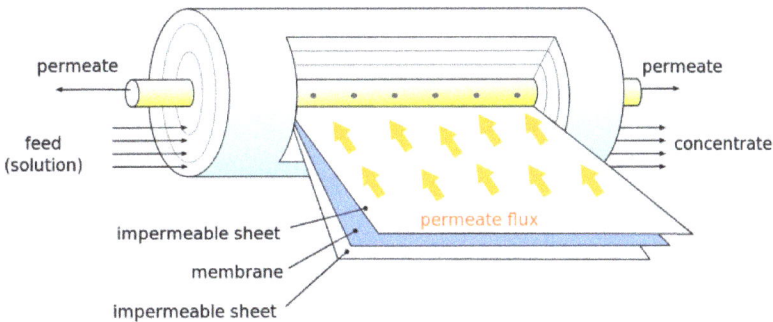

Spiral wound membrane module

Disc tube module is using a cross-flow geometry, and consists of a pressure tube and hydraulic discs, which are held by a central tension rod, and membrane cushions that lie between two discs.

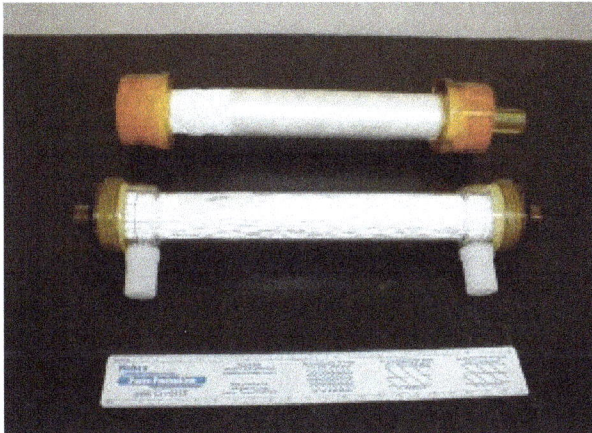

Hollow fiber membrane module

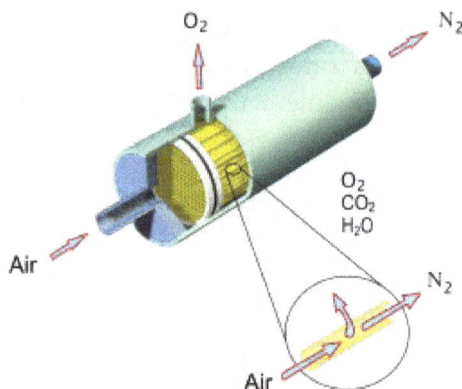

Separation of air into oxygen and nitrogen through a membrane

Membrane Performance and Governing Equations

The selection of synthetic membranes for a targeted separation process is usually based on few requirements. Membranes have to provide enough mass transfer area to process large amounts of feed stream. The selected membrane has to have high selectivity (rejection) properties for certain particles; it has to resist fouling and to have high mechanical stability. It also needs to be reproducible and to have low manufacturing costs. The main modeling equation for the dead-end filtration at constant pressure drop is represented by Darcy's law:

$$\frac{dV_p}{dt} = Q = \frac{\Delta p}{\mu} A\left(\frac{1}{R_m + R}\right)$$

where V_p and Q are the volume of the permeate and its volumetric flow rate respectively (proportional to same characteristics of the feed flow), μ is dynamic viscosity of permeating fluid, A is membrane area, R_m and R are the respective resistances of membrane and growing deposit of the foulants. R_m can be interpreted as a membrane resistance to the solvent (water) permeation. This resistance is a membrane intrinsic property and is expected to be fairly constant and independent of the driving force, Δp. R is related to the type of membrane foulant, its concentration in the filtering solution, and the nature of foulant-membrane interactions. Darcy's law allows for calculation of the membrane area for a targeted separation at given conditions. The solute sieving coefficient is defined by the equation:

$$S = \frac{C_p}{C_f}$$

where C_f and C_p are the solute concentrations in feed and permeate respectively. Hydraulic permeability is defined as the inverse of resistance and is represented by the equation:

$$L_p = \frac{J}{\Delta p}$$

where J is the permeate flux which is the volumetric flow rate per unit of membrane area. The solute sieving coefficient and hydraulic permeability allow the quick assessment of the synthetic membrane performance.

Membrane Separation Processes

Membrane separation processes have a very important role in the separation industry. Nevertheless, they were not considered technically important until the mid-1970s. Membrane separation processes differ based on separation mechanisms and size of the separated particles. The widely used membrane processes include microfiltration, ultrafiltration, nanofiltration, reverse osmosis, electrolysis, dialysis, electrodialysis, gas separation, vapor permeation, pervaporation, membrane distillation, and membrane contactors. All processes except for pervaporation involve no phase change. All processes except (electro)dialysis are pressure driven. Microfiltration and ultrafiltration is widely used in food and beverage processing (beer microfiltration, apple juice ultrafiltration), biotechnological applications and pharmaceutical industry (antibiotic production, protein purification), water purification and wastewater treatment, the microelectronics industry, and others. Nanofiltration and reverse osmosis membranes are mainly used for water purification purposes. Dense membranes are utilized for gas separations (removal of CO_2 from natural gas, separating N_2 from air, organic vapor removal from air or a nitrogen stream) and sometimes in membrane distillation. The later process helps in the separation of azeotropic compositions reducing the costs of distillation processes.

Cut-offs of different liquid filtration techniques							
Micrometer logarithmic scaled	0,001	0,01	0,1	1	10	100	1000
Angstroms logarithmic scaled	1	10	100	1000	10⁴	10⁵	10⁶	10⁷
Molecular weight (Dextran in kD)		0,5	50	7.000				

Ranges of membrane based separations

Pore Size and Selectivity

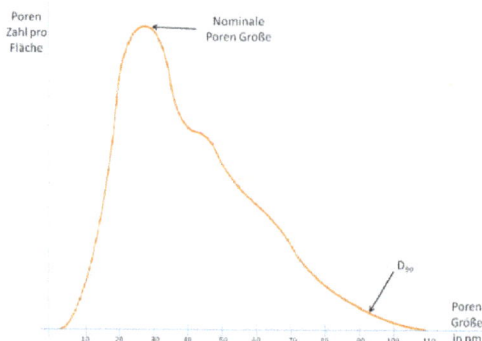

The pore distribution of a fictitious ultrafiltration membrane with the nominal pore size and the D_{90}

The pore sizes of technical membranes are specified differently depending on the manufacturer. One common distinction is by *nominal pore size*. It describes the maximum pore size distribution and gives only vague information about the retention capacity of a membrane. The exclusion limit or "cut-off" of the membrane is usually specified in the form of *NMWC* (nominal molecular weight cut-off, or *MWCO*, molecular weight cut off, with units in Dalton). It is defined as the minimum molecular weight of a globular molecule that is retained to 90% by the membrane. The cut-off, depending on the method, can by converted to so-called D_{90}, which is then expressed in a metric unit. In practice the MWCO of the membrane should be at least 20% lower than the molecular weight of the molecule that is to be separated.

Filter membranes are divided into four classes according to pore size:

Pore size	Molecular mass	Process	Filtration	Removal of
> 10		"Classic" filter		
> 0.1 µm	> 5000 kDa	microfiltration	< 2 bar	larger bacteria, yeast, particles
100-2 nm	5-5000 kDa	ultrafiltration	1-10 bar	bacteria, macromolecules, proteins, larger viruses
2-1 nm	0.1-5 kDa	nanofiltration	3-20 bar	viruses, 2- valent ions
< 1 nm	< 100 Da	reverse osmosis	10-80 bar	salts, small organic molecules

The form and shape of the membrane pores are highly dependent on the manufacturing process and are often difficult to specify. Therefore, for characterization, test filtrations are carried out and the pore diameter refers to the diameter of the smallest particles which could not pass through the membrane.

The rejection can be determined in various ways and provides an indirect measurement of the pore size. One possibility is the filtration of macromolecules (often dextran, polyethylene glycol or albumin), another is measurement of the cut-off by gel permeation chromatography. These methods are used mainly to measure membranes for ultrafiltration applications. Another testing method is the filtration of particles with defined size and their measurement with a particle sizer or by laser induced breakdown spectroscopy (LIBS). A vivid characterization is to measure the rejection of dextran blue or other colored molecules. The retention of bacteriophage and bacteria, the so-called "bacteriachallenge test", can also provide information about the pore size.

Nominal pore size	micro-organism	ATCC root number
0.1 µm	Acholeplasma laidlawii	23206
0.3 µm	Bacillus subtilis spores	82
0.5 µm	Pseudomonas diminuta	19146
0.45 µm	Serratia marcescens	14756
0.65 µm	Lactobacillus brevis	

To determine the pore diameter, physical methods such as porosimetry (mercury, liquid-liquid porosimetry and Bubble Point Test) are also used, but a certain form of the pores (such as cylindrical or concatenated spherical holes) is assumed. Such methods are used for membranes whose pore geometry does not match the ideal, and we get "nominal" pore diameter, which characterizes the membrane, but does not necessarily reflect its actual filtration behavior and selectivity.

The selectivity is highly dependent on the separation process, the composition of the membrane and its electrochemical properties in addition to the pore size. With high selectivity, isotopes can be enriched (uranium enrichment) in nuclear engineering or industrial gases like nitrogen can be recovered (gas separation). Ideally, even racemics can be enriched with a suitable membrane.

When choosing membranes selectivity has priority over a high permeability, as low flows can easily be offset by increasing the filter surface with a modular structure. In gas phase filtration different deposition mechanisms are operative, so that particles having sizes below the pore size of the membrane can be retained as well.

Field Flow Fractionation

Field-flow fractionation, abbreviated FFF, is a separation technique where a field is applied to a fluid suspension or solution pumped through a long and narrow channel, perpendicular to the direction of flow, to cause separation of the particles present in the fluid, depending on their differing "mobilities" under the force exerted by the field. It was invented and first reported by J. Calvin Giddings. The method of FFF is unique to other separation techniques due to the fact that it can separate materials over a wide colloidal size range while maintaining high resolution. Although FFF is an extremely versatile technique, there is no "one size fits all" method for all applications.

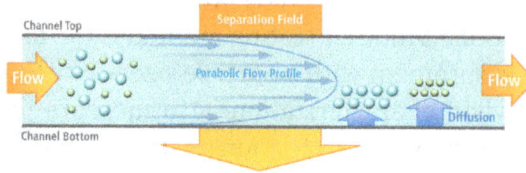

Flow field-flow fractionation (AF4) channel cross section, where the rate of laminar flow within the channel is not uniform. It travels in a parabolic pattern with the speed of the flow, increasing towards the centre of the channel and decreasing towards the sides.

In field-flow fractionation the *field* can be asymmetrical *flow* through a semi-permeable membrane, gravitational, centrifugal, thermal-gradient, electrical, magnetic etc. In all cases, the separation mechanism is born from differences in particle mobility (electrophoretic, when the field is a DC electric field causing a transverse electric current flow) under the forces of the field, *in equilibrium with* the forces of diffusion: an often-parabolic laminar-flow-velocity profile in the channel determines the velocity of a particular particle, based on its equilibrium position from the wall of the channel. The ratio of the velocity of a species of particle to the average velocity of the fluid is called the *retention ratio*.

Fundamental Principles

Field flow fractionation is based on laminar flow of particles in a solution. These sample components will change levels and speed based on their size/mass. Since these components will be travelling at different speeds, separation occurs. A simplified explanation of the setup is as follows. The sample separation occurs in a thin, ribbon-like, channel in which there is an inlet flow and a perpendicular field flow. The inlet flow is where the carrier liquid is pumped into the channel and it creates a parabolic flow profile and it propels the sample towards the outlet of the channel.

Relating Force (F) to Retention Time (TR)

The relationship between the separative force field and retention time can be illustrated from first principles. Consider two particle populations within the FFF channel. The cross field drives both particle clouds towards the bottom "accumulation" wall. Opposing this force field is the particles natural diffusion, or Brownian motion, which produces a counter acting motion. When these two transport process reach equilibrium the particle concentration c approaches the exponential function of elevation x above the accumulation wall as illustrated in equation 1.

$$c = c_0 e^{\frac{-x}{l}}$$

l represents the characteristic elevation of the particle cloud. This relates to the height that the particle group can reach within the channel and only when the value for l is different for either group will separation occur. The l of each component can be related to the force applied on each individual particle.

$$l = \frac{kT}{F}$$

Where k is the Boltzmann constant, T is absolute pressure and F is the force exerted on a single particle by the cross flow. This shows how the characteristic elevation value is inversely dependent to the Force applied. Therefore, F governs the separation process. Hence, by varying the field strength the separation can be controlled to achieve optimal levels. The velocity V of a cloud of molecules is simply the average velocity of an exponential distribution embedded in a parabolic flow profile. Retention time, tr can be written as:

$$t_r = \frac{L}{V}$$

Where L is the channel length. Subsequently, the retention time can be written as:

$$t_r/t^o = w/6l \lfloor coth \; w/2l- 2l/w \rfloor^{-1}$$

Where to is the void time (emergence of a non-retained tracer) and w is the sample thickness. Substituting in kT/F in place of l illustrates the retention time with respect to the cross force applied.

$$t_r/t^o = Fw/6kT \lfloor coth \; Fw/2kT- 2kT/Fw \rfloor^{-1}$$

For an efficient operation the channel thickness value w far exceeds l. When this is the case the term in the brackets approaches unity. Therefore, equation 5 can be approximated as:

$$t_r/t^o = w/6l = Fw/6kT$$

Thus tr is roughly proportional to F. The separation of particle bands X and Y, represented by the finite increment Δtr in their retention times, is achieved only if the force increment ΔF between them is sufficient. A differential in force of only 10–16 N is required for this to be the case. The magnitude of F and ΔF depend on particle properties, field strength and the type of field. This allows for variations and specialisations for the technique. From this basic principle many forms of FFF have evolved varying by the nature of the separative force applied and the range in molecule size to which they are targeted.

Fractogram

Centrifugal FFF can be separated by particle density and particle size. With identical sized gold and silver nanoparticles can be separated into two peaks, according to differences in density in the gold and silver nanoparticles.

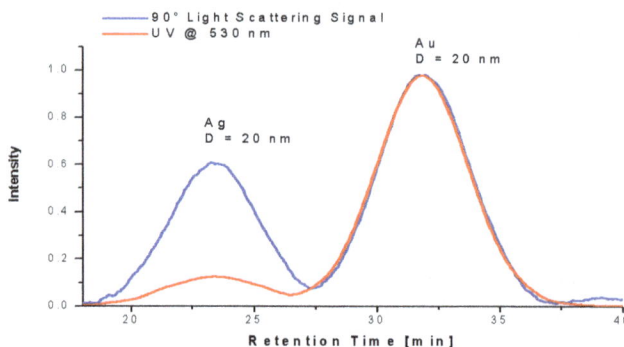

A graph of a detection signal vs. time, derived from an FFF process, in which various substances present in a fluid get separated based on their flow velocities under some applied external field, such as an flow, centrifugal, thermal or electric field.

Often these substances are various particles initially suspended in a small volume of a liquid buffer and pushed along a fractionation channel by more of the pure buffer. The varying velocities of a particular species of particles may be due to its size, its mass, and/or its distance from the walls of a channel with non-uniform flow-velocity. The presence of different species in a sample can thus be identified through detection of a common property at some distance down the long channel, and by the resulting fractogram indicating the presence of the various species by peaks, due to the different times of arrival characteristic of each species and its physical and chemical properties.

In an electrical FFF, an electric field controls the velocity by controlling the lateral position of either a charged (having electrophoretic mobility) or polarized (being levitated in a non-uniform field) species in a capillary channel with a hydrodynamically parabolic flow-velocity profile, meaning that the velocity of the pumped fluid is highest midway between the walls of the channel and it monotonically decays to a minimum of zero at the wall surface.

Forms of FFF

Most techniques available today are advances on those originally created by Prof. Giddings nearly 4 decades ago.

Flow FFF

Of these techniques Flow FFF was the first to be offered commercially. Flow FFF separates particles based on size, independent of density and can measure macromolecules in the range of 1 nm to 1 μm. In this respect it is the most sensitive FFF technique available. The cross flow in Flow FFF enters through a porous frit at the top of the channel, exiting through a semi-permeable membrane outlet frit on the accumulation wall (i.e. the bottom wall).

Hollow Fiber Flow FFF

Hollow fiber flow FFF (HF5) has been pioneered by H.L. Lee, J.F.G. Reis and E. N. Lightfoot. HF5 has been applied towards the analysis of lattices and other macromolecules. HF5 was the first form of flow FFF (1974) to be developed. Flat membranes soon outperformed hollow fibers and forced HF5 into obscurity. One of the drawbacks of HF5 is the availability of the membranes with uniform pore sizes. There are different kinds of ceramic and polymeric hollow fiber membranes used in practice.

Asymmetric Flow FFF (AF4)

Asymmetric Flow FFF (AF4), on the other hand, has only one semi-permeable membrane on the bottom wall of the channel. The cross flow is, therefore, created by the carrier liquid exiting the bottom of the channel. This offers an extremely gentle separation and an "ultra-broad" separation range. High Temperature Asymmetric Flow Field-Flow Fractionation is the most advanced technology for the separation of high and ultra-high molar mass polymers, macromolecules and nanoparticles in the size range.

Thermal FFF

Thermal FFF, as the name suggests, establishes a separation force by applying a temperature gradient to the channel. The top channel wall is heated and the bottom wall is cooled driving polymers and particles towards the cold wall by thermal diffusion. Thermal FFF was developed as a technique for separating synthetic polymers in organic solvents. Thermal FFF is unique amongst FFF techniques in that it can separate macromolecules by both molar mass and chemical composition, allowing for the separation of polymer fractions with the same molecular weight. Today this technique is ideally suited for the characterization of polymers, gels and nanoparticles.

Split Flow Thin-cell Fractionation (SPLITT)

Split flow thin-cell fractionation (SPLITT) is a special preparative FFF technique, using gravity for separation of μm-sized particles on a continuous basis. SPLITT is performed by pumping the sample containing liquid into the top inlet at the start of the channel, whilst simultaneously pumping a carrier liquid into the bottom inlet. By controlling the flow rate ratios of the two inlet streams and two outlet streams, the separation can be controlled and the sample separated into two distinct sized fractions. The use of gravity alone as the separating force makes SPLITT the least sensitive FFF technique, limited to particles above 1 μm.

Centrifugal FFF

With further developments in sedimentation FFF, this has led to the development of a new technique, centrifugal FFF, wherein the separation field is supplied via a centrifugal force. The channel takes the form of a ring, which spins at 4900 rpm. The flow and sample are pumped into the chamber and the mixture is centrifuged, allowing the operator to resolve the particles by size and density. The advantage of centrifugal FFF lies in the broad range of samples and high resolution that can be achieved by varying the speed and force applied.

The unique advantage presented by centrifugal FFF comes from the techniques capability for high resolution. The first commercial centrifugal FFF instrument was introduced by Postnova Analytics is the CF2000, incorporating the unique feature of separating particles by dynamic diffusion on the basis of both particle size and density. This allows for the separation of particles with only a 5% difference in size. Centrifugal FFF has the advantage that molecules can be separated by particle density, rather than just particle size. In this instance, two identically sized gold and silver nanoparticles can be separated into two peaks, according to differences in density in the gold and silver nanoparticles, separated with the Centrifugal FFF Postnova CF2000 instrument with the detection by Dynamic Light Scattering (DLS).

In AF4 separations, the ratio of mass to time is 1:1. With the addition of the third parameter of density to centrifugal fractionation, this produces a ratio more akin to mass:time to the power of three. This results in a significantly larger distinction between peaks and result in a greatly improved resolution. This can be particularly useful for novel products, such as composite materials and coated polymers containing nanoparticles, particles which may not vary in size but do vary in density. In this way two identically sized particles can still be separated into two peaks, providing that the density is different.

Theoretical Plate

A theoretical plate in many separation processes is a hypothetical zone or stage in which two phases, such as the liquid and vapor phases of a substance, establish an equilibrium with each other. Such equilibrium stages may also be referred to as an equilibrium stage, ideal stage, or a theoretical tray. The performance of many separation processes depends on having a series of equilibrium stages and is enhanced by providing more such stages. In other words, having more theoretical plates increases the efficiency of the separation process be it either a distillation, absorption, chromatographic, adsorption or similar process.

Applications

The concept of theoretical plates and trays or equilibrium stages is used in the design of many different types of separation.

Distillation Columns

The concept of theoretical plates in designing distillation processes has been discussed in many reference texts. Any physical device that provides good contact between the vapor and liquid phases present in industrial-scale distillation columns or laboratory-scale glassware distillation columns constitutes a "plate" or "tray". Since an actual, physical plate is rarely a 100% efficient equilibrium stage, the number of actual plates is more than the required theoretical plates.

$$N_a = \frac{N_t}{E}$$

where N_a is the number of actual, physical plates or trays, N_t is the number of theoretical plates or trays and E is the plate or tray efficiency.

So-called bubble-cap or valve-cap trays are examples of the vapor and liquid contact devices used in industrial distillation columns. Another example of vapor and liquid contact devices are the spikes in laboratory Vigreux fractionating columns.

The trays or plates used in industrial distillation columns are fabricated of circular steel plates and usually installed inside the column at intervals of about 60 to 75 cm (24 to 30 inches) up the height of the column. That spacing is chosen primarily for ease of installation and ease of access for future repair or maintenance.

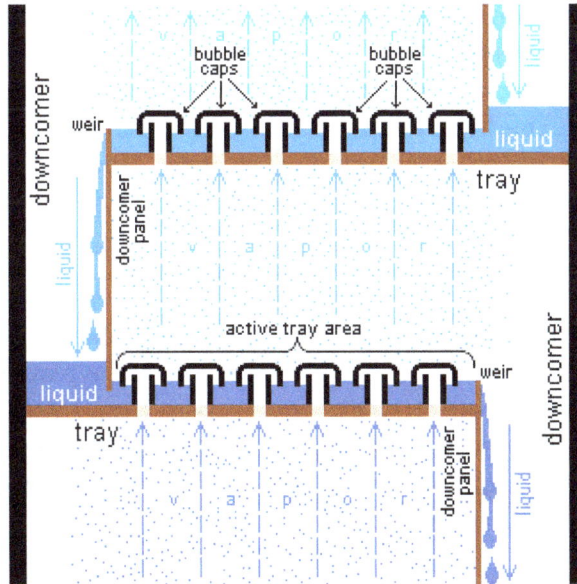

Typical bubble cap trays used in industrial distillation columns

An example of a very simple tray is a perforated tray. The desired contacting between vapor and liquid occurs as the vapor, flowing upwards through the perforations, comes into contact with the liquid flowing downwards through the perforations. In current modern practice, as shown in the adjacent diagram, better contacting is achieved by installing bubble-caps or valve caps at each perforation to promote the formation of vapor bubbles flowing through a thin layer of liquid maintained by a weir on each tray.

To design a distillation unit or a similar chemical process, the number of theoretical trays or plates (that is, hypothetical equilibrium stages), N_t, required in the process should be determined, taking into account a likely range of feedstock composition and the desired degree of separation of the components in the output fractions. In industrial continuous fractionating columns, N_t is determined by starting at either the top or bottom of the column and calculating material balances, heat balances and equilibrium flash vaporizations for each of the succession of equilibrium stages until the desired end product composition is achieved. The calculation process requires the availability of a great deal of vapor–liquid equilibrium data for the components present in the distillation feed, and the calculation procedure is very complex.

In an industrial distillation column, the N_t required to achieve a given separation also depends upon the amount of reflux used. Using more reflux decreases the number of plates required and using less reflux increases the number of plates required. Hence, the calculation of N_t is usually repeated at various reflux rates. N_t is then divided by the tray efficiency, E, to determine the actual number of trays or physical plates, N_a, needed in the separating column. The final design choice of the number of trays to be installed in an industrial distillation column is then selected based upon an economic balance between the cost of additional trays and the cost of using a higher reflux rate.

There is a very important distinction between the theoretical plate terminology used in discussing conventional distillation trays and the theoretical plate terminology used in the discussions below of packed bed distillation or absorption or in chromatography or other applications. The theoretical plate in conventional distillation trays has no "height". It is simply a hypothetical equilibrium stage. However, the theoretical plate in packed beds, chromatography and other applications is defined as having a height.

Distillation and Absorption Packed Beds

Distillation and absorption separation processes using packed beds for vapor and liquid contacting have an equivalent concept referred to as the plate height or the height equivalent to a theoretical plate (HETP). HETP arises from the same concept of equilibrium stages as does the theoretical plate and is numerically equal to the absorption bed length divided by the number of theoretical plates in the absorption bed (and in practice is measured in this way).

$$N_t = \frac{H}{\text{HETP}}$$

where N_t is the number of theoretical plates (also called the "plate count"), H is the total bed height and HETP is the height equivalent to a theoretical plate.

The material in packed beds can either be random dumped packing (1-3" wide) such as Raschig rings or structured sheet metal. Liquids tend to wet the surface of the packing and the vapors contact the wetted surface, where mass transfer occurs.

Chromatographic Processes

The theoretical plate concept was also adapted for chromatographic processes by Martin and Synge. The IUPAC's Gold Book provides a definition of the number of theoretical plates in a chromatography column.

The same equation applies in chromatography processes as for the packed bed processes, namely:

$$N_t = \frac{H}{\mathrm{HETP}}$$

In chromatography, the HETP may also be calculated with the Van Deemter equation.

Other Applications

The concept of theoretical plates or trays applies to other processes as well, such as capillary electrophoresis and some types of adsorption.

References

- Wilson, Ian D.; Adlard, Edward R.; Cooke, Michael; et al., eds. (2000). Encyclopedia of separation science. San Diego: Academic Press. ISBN 978-0-12-226770-3.

- McMurry, John (2011). Organic chemistry: with biological applications (2nd ed.). Belmont, CA: Brooks/Cole. p. 395. ISBN 9780495391470.

- Harwood, Laurence M.; Moody, Christopher J. (1989). Experimental organic chemistry: Principles and Practice (Illustrated ed.). WileyBlackwell. pp. 180–185. ISBN 978-0-632-02017-1.

- Anfinsen, Christian B.; Edsall, John Tileston; Richards, Frederic Middlebrook, eds. (1976). Advances in Protein Chemistry. pp. 6–7. ISBN 978-0-12-034230-3.

- Bailon, Pascal; Ehrlich, George K.; Fung, Wen-Jian and Berthold, Wolfgang (2000) An Overview of Affinity Chromatography, Humana Press. ISBN 978-0-89603-694-9.

- Bryan H. Bunch; Alexander Hellemans (2004). The History of Science and Technology. Houghton Mifflin Harcourt. p. 88. ISBN 0-618-22123-9.

- Forbes, Robert James (1970). A short history of the art of distillation: from the beginnings up to the death of Cellier Blumenthal. BRILL. pp. 57, 89. ISBN 978-90-04-00617-1. Retrieved 29 June 2010.

- D. F. Othmer (1982) "Distillation – Some Steps in its Development", in W. F. Furter (ed) A Century of Chemical Engineering ISBN 0-306-40895-3

- Perry, Robert H.; Green, Don W. (1984). Perry's Chemical Engineers' Handbook (6th ed.). McGraw-Hill. ISBN 0-07-049479-7.

- Gavin Towler & R K Sinnott (2007). Chemical Engineering Design: Principles, Practice and Economics of Plant and Process Design. Butterworth-Heinemann. ISBN 0-7506-8423-2.

- Perry, Robert H. & Green, Don W. (1984). Perry's Chemical Engineers' Handbook (6th ed.). McGraw-Hill. ISBN 0-07-049479-7.

Major Aspects of Chemical Engineering

Transport phenomena, Stefan tube, chemical reactor, process simulation and chemical kinetics are some of the major aspects of chemical engineering. Transport phenomena deals with the exchange of mass and energy between systems whereas Stefan tubes are devices used for measuring diffusion coefficients. Chemical engineering is best understood in confluence with the major topics listed in the following text.

Transport Phenomena

In engineering, physics and chemistry, the study of transport phenomena concerns the exchange of mass, energy, and momentum between observed and studied systems. While it draws from fields as diverse as continuum mechanics and thermodynamics, it places a heavy emphasis on the commonalities between the topics covered. Mass, momentum, and heat transport all share a very similar mathematical framework, and the parallels between them are exploited in the study of transport phenomena to draw deep mathematical connections that often provide very useful tools in the analysis of one field that are directly derived from the others. While it draws its theoretical foundation from principles in a number of fields, most of the fundamental transport theory is a restatement of basic conservation laws. The fundamental analyses in all three subfields of mass, heat, and momentum transfer are often grounded in the simple principle that the sum total of the quantities being studied must be conserved by the system and its environment. Thus, the different phenomena that lead to transport are each considered individually with the knowledge that the sum of their contributions must equal zero. This principle is useful for calculating many relevant quantities. For example, in fluid mechanics, a common use of transport analysis is to determine the velocity profile of a fluid flowing through a rigid volume.

Transport phenomena are ubiquitous throughout the engineering disciplines. Some of the most common examples of transport analysis in engineering are seen in the fields of process, chemical, biological, and mechanical engineering, but the subject is a fundamental component of the curriculum in all disciplines involved in any way with fluid mechanics, heat transfer, and mass transfer. It is now considered to be a part of the engineering discipline as much as thermodynamics, mechanics, and electromagnetism.

Transport phenomena encompass all agents of physical change in the universe. Moreover, they are considered to be fundamental building blocks which developed the universe, and which is responsible for the success of all life on earth. However, the scope here is limited to the relationship of transport phenomena to artificial engineered systems.

Overview

In physics, transport phenomena are all irreversible processes of statistical nature stemming from the random continuous motion of molecules, mostly observed in fluids. Every aspect of transport phenomena is grounded in two primary concepts : the conservation laws, and the constitutive equations. The conservation laws, which in the context of transport phenomena are formulated as continuity equations, describe how the quantity being studied must be conserved. The constitutive equations describe how the quantity in question responds to various stimuli via transport. Prominent examples include Fourier's Law of Heat Conduction and the Navier-Stokes equations, which describe, respectively, the response of heat flux to temperature gradients and the relationship between fluid flux and the forces applied to the fluid. These equations also demonstrate the deep connection between transport phenomena and thermodynamics, a connection that explains why transport phenomena are irreversible. Almost all of these physical phenomena ultimately involve systems seeking their lowest energy state in keeping with the principle of minimum energy. As they approach this state, they tend to achieve true thermodynamic equilibrium, at which point there are no longer any driving forces in the system and transport ceases. The various aspects of such equilibrium are directly connected to a specific transport: heat transfer is the system's attempt to achieve thermal equilibrium with its environment, just as mass and momentum transport move the system towards chemical and mechanical equilibrium.

Examples of transport processes include heat conduction (energy transfer), fluid flow (momentum transfer), molecular diffusion (mass transfer), radiation and electric charge transfer in semiconductors.

Transport phenomena have wide application. For example, in solid state physics, the motion and interaction of electrons, holes and phonons are studied under "transport phenomena". Another example is in biomedical engineering, where some transport phenomena of interest are thermoregulation, perfusion, and microfluidics. In chemical engineering, transport phenomena are studied in reactor design, analysis of molecular or diffusive transport mechanisms, and metallurgy.

The transport of mass, energy, and momentum can be affected by the presence of external sources:

- An odor dissipates more slowly (and may intensify) when the source of the odor remains present.

- The rate of cooling of a solid that is conducting heat depends on whether a heat source is applied.

- The gravitational force acting on a rain drop counteracts the resistance or drag imparted by the surrounding air.

Commonalities Among Phenomena

An important principle in the study of transport phenomena is analogy between phenomena.

Diffusion

There are some notable similarities in equations for momentum, energy, and mass transfer which can all be transported by diffusion, as illustrated by the following examples:

- Mass: the spreading and dissipation of odors in air is an example of mass diffusion.

- Energy: the conduction of heat in a solid material is an example of heat diffusion.

- Momentum: the drag experienced by a rain drop as it falls in the atmosphere is an example of momentum diffusion (the rain drop loses momentum to the surrounding air through viscous stresses and decelerates).

The molecular transfer equations of Newton's law for fluid momentum, Fourier's law for heat, and Fick's law for mass are very similar. One can convert from one transfer coefficient to another in order to compare all three different transport phenomena.

A great deal of effort has been devoted in the literature to developing analogies among these three transport processes for turbulent transfer so as to allow prediction of one from any of the others. The Reynolds analogy assumes that the turbulent diffusivities are all equal and that the molecular diffusivities of momentum (μ/ρ) and mass (D_{AB}) are negligible compared to the turbulent diffusivities. When liquids are present and/or drag is present, the analogy is not valid. Other analogies, such as von Karman's and Prandtl's, usually result in poor relations.

The most successful and most widely used analogy is the Chilton and Colburn J-factor analogy. This analogy is based on experimental data for gases and liquids in both the laminar and turbulent regimes. Although it is based on experimental data, it can be shown to satisfy the exact solution derived from laminar flow over a flat plate. All of this information is used to predict transfer of mass.

Onsager Reciprocal Relations

In fluid systems described in terms of temperature, matter density, and pressure, it is known that temperature differences lead to heat flows from the warmer to the cold-er parts of the system; similarly, pressure differences will lead to matter flow from high-pressure to low-pressure regions (a "reciprocal relation"). What is remarkable is the observation that, when both pressure and temperature vary, temperature differences at constant pressure can cause matter flow (as in convection) and pressure differences at constant temperature can cause heat flow. Perhaps surprisingly, the heat flow per unit of pressure difference and the density (matter) flow per unit of temperature difference are equal.

This equality was shown to be necessary by Lars Onsager using statistical mechanics as a consequence of the time reversibility of microscopic dynamics. The theory developed by Onsager is much more general than this example and capable of treating more than two thermodynamic forces at once.

Momentum Transfer

In momentum transfer, the fluid is treated as a continuous distribution of matter. The study of momentum transfer, or fluid mechanics can be divided into two branches: fluid statics (fluids at rest), and fluid dynamics (fluids in motion). When a fluid is flowing in the x direction parallel to a solid surface, the fluid has x-directed momentum, and its concentration is $v_x \rho$. By random diffusion of molecules there is an exchange of molecules in the z direction. Hence the x-directed momentum has been transferred in the z-direction from the faster- to the slower-moving layer. The equation for momentum transport is Newton's Law of Viscosity written as follows:

$$\tau_{zx} = -v \frac{\partial \rho v_x}{\partial z}$$

where τ_{zx} is the flux of x-directed momentum in the z direction, v is μ/ρ, the momentum diffusivity, z is the distance of transport or diffusion, ρ is the density, and μ is the viscosity. Newtons Law is the simplest relationship between the flux of momentum and the velocity gradient.

Mass Transfer

When a system contains two or more components whose concentration vary from point to point, there is a natural tendency for mass to be transferred, minimizing any concentration difference within the system. Mass Transfer in a system is governed by Fick's First Law: 'Diffusion flux from higher concentration to lower concentration is proportional to the gradient of the concentration of the substance and the diffusivity of the substance in the medium.' Mass transfer can take place due to different driving forces. Some of them are:

- Mass can be transferred by the action of a pressure gradient(pressure diffusion)

- Forced diffusion occurs because of the action of some external force

- Diffusion can be caused by temperature gradients (thermal diffusion)

- Diffusion can be caused by differences in chemical potential

This can be compared to Fick's Law of Diffusion:

$$J_{Ay} = -D_{AB} \frac{\partial Ca}{\partial y}$$

where D is the diffusivity constant.

Energy Transfer

All processes in engineering involve the transfer of energy. Some examples are the heating and cooling of process streams, phase changes, distillations, etc. The basic principle is the first law of thermodynamics which is expressed as follows for a static system:

$$q = -k \frac{dT}{dx}$$

The net flux of energy through a system equals the conductivity times the rate of change of temperature with respect to position.

For other systems that involve either turbulent flow, complex geometries or difficult boundary conditions another equation would be easier to use:

$$Q = h \cdot A \cdot \Delta T$$

where A is the surface area, : ΔT is the temperature driving force, Q is the heat flow per unit time, and h is the heat transfer coefficient.

Within heat transfer, two types of convection can occur:

Forced convection can occur in both laminar and turbulent flow. In the situation of laminar flow in circular tubes, several dimensionless numbers are used such as Nusselt number, Reynolds number, and Prandtl. The commonly used equation is:

$$Nu_a = \frac{h_a D}{k}$$

Natural or free convection is a function of Grashof and Prandtl numbers. The complexities of free convection heat transfer make it necessary to mainly use empirical relations from experimental data.

Heat transfer is analyzed in packed beds, reactors and heat exchangers.

Stefan Tube

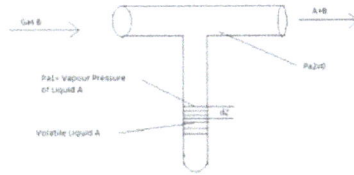

A schematic of a stefan tube. Note that it is not necessary for there to be a cross-wise horizontal tube at the top.

In chemical engineering, a Stefan tube is a device that was devised by Josef Stefan in 1874. It is often used for measuring diffusion coefficients. It comprises a vertical tube, over the top of which a gas flows and at the bottom of which is a pool of volatile liquid that is maintained in a constant-temperature bath. The liquid in the pool evaporates, diffuses through the gas above it in the tube, and is carried away by the gas flow over the tube mouth at the top. One then measures the fall in the level of the liquid in the tube.

The tube conventionally has a narrow diameter, in order to suppress convection.

The way that a Stefan tube is modelled, mathematically, is very similar to how one can model the diffusion of perfume fragrance molecules from (say) a drop of perfume on skin or clothes, evaporating up through the air to a person's nose. There are some differences between the models. However, they turn out to have little effect on results at highly dilute vapour concentrations.

Analysis

In the analysis of the system, various assumptions are made. The liquid, conventionally denoted A, is neither soluble in the gas in the tube, conventionally denoted B, nor reacts with it. The decrease in volume of the liquid A and increase in volume of the gas B over time can be ignored for the purposes of solving the equations that describe the behaviour, and an assumption can be made that the instantaneous flux at any time is the steady state value. There are no radial or circumferential components to the concentration gradients, resulting from convection or turbulence caused by excessively vigorous flow at the upper mouth of the tube, and the diffusion can thus be treated as a simple one-dimensional flow in the vertical direction. The mole fraction of A at the upper mouth of the tube is zero, as a consequence of the gas flow. At the interface between A and B the flux of B is zero (because it is insoluble in A) and the mole fraction is the equilibrium value.

The flux of B, denoted N_B, is thus zero throughout the tube, its diffusive flux downward (along its concentration gradient) is balanced by its convective flux upward caused by A.

Applying these assumptions, the system can be modelled using Fick's laws of diffusion or as Maxwell–Stefan diffusion.

Chemical Reactor

In chemical engineering, chemical reactors are vessels designed to contain chemical reactions. Also referred to as a reaction vessel, the reactants contained are substances that change form after a chemical reaction. One example is a pressure reactor. The design of a chemical reactor deals with multiple aspects of chemical engineering. Chemical engineers design reactors to maximize net present value for the given reaction. Designers ensure that the reaction proceeds with the highest efficiency towards the desired output product, producing the highest yield of product while requiring the least amount of money to purchase and operate. Normal operating expenses include energy input, energy removal, raw material costs, labor, etc. Energy changes can come in the form of heating or cooling, pumping to increase pressure, frictional pressure loss (such as pressure drop across a 90° elbow or an orifice plate) or agitation.

Chemical reaction engineering is the branch of chemical engineering which deals with chemical reactors and their design, especially by application of chemical kinetics to industrial systems.

Overview

Cut-away view of a stirred-tank chemical reactor with a cooling jacket

Chemical reactor with half coils wrapped around it

There are a couple of main basic vessel types:

- A tank
- A pipe or tubular reactor (laminar flow reactor(LFR))

Both types can be used as continuous reactors or batch reactors, and either may accommodate one or more solids (reagents, catalyst, or inert materials), but the reagents and products are typically fluids. Most commonly, reactors are run at steady-state, but can also be operated in a transient state. When a reactor is first brought into operation (after maintenance or inoperation) it would be considered to be in a transient state, where key process variables change with time.

There are three main basic models used to estimate the most important process variables of different chemical reactors:

- *batch reactor* model (batch),
- *continuous stirred-tank reactor* model (CSTR), and
- *plug flow reactor* model (PFR).

Furthermore, catalytic reactors require separate treatment, whether they are batch, CST, or PF reactors, as the many assumptions of the simpler models are not valid.

Key process variables include

- Residence time (τ, lower case Greek tau)
- Volume (V)
- Temperature (T)
- Pressure (P)
- Concentrations of chemical species (C_1, C_2, C_3, ... C_n)
- Heat transfer coefficients (h, U)
- Chemical reactor manufacturer (h, U)

A chemical reactor, typically tubular reactor, could be a packed bed. Typical packed bed reactors consist of a chamber, such as a tube or channel that contains catalyst particles or pellets, and a liquid that flows through the catalyst. A chemical reactor may also be a fluidized bed.

Chemical reactions occurring in a reactor may be exothermic, meaning giving off heat, or endothermic, meaning absorbing heat. A chemical reactor vessel may have a cooling or heating jacket or cooling or heating coils (tubes) wrapped around the outside of its vessel wall to cool down or heat up the contents.

Types

Batch Reactor

The simplest type of reactor is a batch reactor. Materials are charged in a batch reactor, and the reaction proceeds with time. A batch reactor does not reach a steady state, and control of temperature, pressure and volume is necessary. Thus, a batch reactor has ports for sensors and material input and output. Batch reactors are used in small-scale production and reactions with biological materials, such as in brewing, pulping and production of enzymes.

CSTR (Continuous Stirred-Tank Reactor)

Checking condition inside the case of a continuous stirred tank reactor (CSTR). Note the impeller (or agitator) blades on the shaft for mixing. Also note the baffle at the bottom of the image which also helps in mixing.

In a CSTR, one or more fluid reagents are introduced into a tank reactor (typically) equipped with an impeller while the reactor effluent is removed. The impeller stirs the reagents to ensure proper mixing. Simply dividing the volume of the tank by the average volumetric flow rate through the tank gives the *space time*, or the average amount of time a discrete quantity of reagent spends inside the tank. Using chemical kinetics, the reaction's expected percent completion can be calculated. Some important aspects of the CSTR:

- At steady-state, the mass flow rate in must equal the mass flow rate out, otherwise the tank will overflow or go empty (transient state). While the reactor is in a transient state the model equation must be derived from the differential mass and energy balances.

- The reaction proceeds at the reaction rate associated with the final (output) concentration, since the concentration is assumed to be homogenous throughout the reactor.

- Often, it is economically beneficial to operate several CSTRs in series. This allows, for example, the first CSTR to operate at a higher reagent concentration and therefore a higher reaction rate. In these cases, the sizes of the reactors may be varied in order to minimize the total capital investment required to implement the process.

- It can be demonstrated that an infinite number of infinitely small CSTRs operating in series would be equivalent to a PFR.

The behavior of a CSTR is often approximated or modeled by that of a Continuous Ideally Stirred-Tank Reactor (CISTR). All calculations performed with CISTRs assume perfect mixing. If the residence time is 5-10 times the mixing time, this approximation is considered valid for engineering purposes. The CISTR model is often used to simplify engineering calculations and can be used to describe research reactors. In practice it can only be approached, particularly in industrial size reactors in which the mixing time may be very large.

A loop reactor is a hybrid type of catalytic reactor that physically resembles a tubular reactor, but operates like a CSTR. The reaction mixture is circulated in a loop of tube, surrounded by a jacket for cooling or heating, and there is a continuous flow of starting material in and product out.

PFR (Plug Flow Reactor)

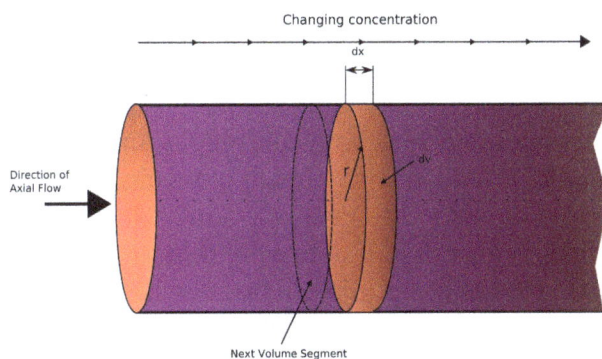

Simple diagram illustrating plug flow reactor model

In a PFR, sometimes called continuous tubular reactor (CTR), one or more fluid reagents are pumped through a pipe or tube. The chemical reaction proceeds as the reagents travel through the PFR. In this type of reactor, the changing reaction rate creates a gradient with respect to distance traversed; at the inlet to the PFR the rate is very high, but as the concentrations of the reagents decrease and the concentration of the product(s) increases the reaction rate slows. Some important aspects of the PFR:

- All calculations performed with PFRs assume no upstream or downstream mixing, as implied by the term "plug flow".

- Reagents may be introduced into the PFR at locations in the reactor other than the inlet. In this way, a higher efficiency may be obtained, or the size and cost of the PFR may be reduced.

- A PFR typically has a higher efficiency than a CSTR of the same volume. That is, given the same space-time (or residence time), a reaction will typically proceed to a higher percentage completion in a PFR than in a CSTR. This is not always true for reversible reactions.

For most chemical reactions of industrial interest, it is impossible for the reaction to proceed to 100% completion. The rate of reaction decreases as the reactants are consumed until the point where the system reaches dynamic equilibrium (no net reaction, or change in chemical species occurs). The equilibrium point for most systems is less than 100% complete. For this reason a separation process, such as distillation, often follows a chemical reactor in order to separate any remaining reagents or byproducts from the desired product. These reagents may sometimes be reused at the beginning of the process, such as in the Haber process. In some cases, very large reactors would be necessary to approach equilibrium, and chemical engineers may choose to separate the partially reacted mixture and recycle the leftover reactants.

Continuous oscillatory baffled reactor (COBR) is a tubular plug flow reactor. The mixing in COBR is achieved by the combination of fluid oscillation and orifice baffles, allowing plug flow to be achieved under laminar flow conditions with the net flow Reynolds number just about 100.

Semi-batch Reactor

A semi-batch reactor is operated with both continuous and batch inputs and outputs. A fermenter, for example, is loaded with a batch of medium and microbes which constantly produces carbon dioxide that must be removed continuously. Analogously, driving a reaction of gas with a liquid is usually difficult, since the gas bubbles off. Therefore, a continuous feed of gas is injected into the batch of a liquid. One chemical reactant is charged to the vessel and a second chemical is added slowly (for instance, to prevent side reactions).

Catalytic Reactor

Although catalytic reactors are often implemented as plug flow reactors, their analysis requires more complicated treatment. The rate of a catalytic reaction is proportional to the amount of catalyst the reagents contact, as well as the concentration of the reactants. With a solid phase catalyst and fluid phase reagents, this is proportional to the exposed area, efficiency of diffusion of reagents in and products out, and efficacy of mixing. Perfect mixing usually cannot be assumed. Furthermore, a catalytic reaction pathway often occurs in multiple steps with intermediates that are chemically bound to the catalyst; and as the chemical binding to the catalyst is also a chemical reaction, it may affect the kinetics. Catalytic reactions often display the so-called *falsified kinetics*, i.e. the apparent kinetics differ from elementary chemical kinetics due to physical transport effects.

The behavior of the catalyst is also a consideration. Particularly in high-temperature petrochemical processes, catalysts are deactivated by sintering, coking, and similar processes.

A common example of a catalytic reactor is the catalytic converter following an engine. However, most petrochemical reactors are catalytic, and are responsible for most of industrial chemical production in the world, with extremely high-volume examples such as sulfuric acid, ammonia, reformate/BTEX (benzene, toluene, ethylbenzene and xylene) and alkylate gasoline blending stock.

Chemical Plant

BASF Chemical Plant Portsmouth Site in the West Norfolk area of Portsmouth, Virginia, United States. The plant is served by the Commonwealth Railway.

A chemical plant is an industrial process plant that manufactures (or otherwise processes) chemicals, usually on a large scale. The general objective of a chemical plant is to create new material wealth via the chemical or biological transformation and or separation of materials. Chemical plants use specialized equipment, units, and technology in the manufacturing process. Other kinds of plants, such as polymer, pharmaceutical, food, and some beverage production facilities, power plants, oil refineries or other refineries, natural gas processing and biochemical plants, water and wastewater treatment, and pollution control equipment use many technologies that have similarities to chemical plant technology such as fluid systems and chemical reactor systems. Some would consider an oil refinery or a pharmaceutical or polymer manufacturer to be effectively a chemical plant.

Petrochemical plants (plants using chemicals from petroleum as a raw material or *feedstock*) are usually located adjacent to an oil refinery to minimize transportation costs for the feedstocks produced by the refinery. Speciality chemical and fine chemical plants are usually much smaller and not as sensitive to location. Tools have been developed for converting a base project cost from one geographic location to another.

Chemical Processes

Chemical plants use chemical processes, which are detailed industrial-scale methods, to transform feedstock chemicals into products. The same chemical process can be used at more than one chemical plant, with possibly differently scaled capacities at each plant. Also, a chemical plant at a site may be constructed to utilize more than one chemical process, for instance to produce multiple products.

A chemical plant commonly has usually large vessels or sections called units or lines that are interconnected by piping or other material-moving equipment which can carry streams of material. Such material streams can include fluids (gas or liquid carried in piping) or sometimes solids or mixtures such as slurries. An overall chemical process is commonly made up of steps called unit operations which occur in the individual units. A raw material going into a chemical process or plant as input to be converted into a product is commonly called a feedstock, or simply feed. In addition to feedstocks for the plant as a whole, an input stream of material to be processed in a particular unit can similarly be considered feed for that unit. Output streams from the plant as a whole are final products and sometimes output streams from individual units may be considered intermediate products for their units. However, final products from one plant may be intermediate chemicals used as feedstock in another plant for further processing. For example, some products from an oil refinery may used as feedstock in petrochemical plants, which may in turn produce feedstocks for pharmaceutical plants.

Either the feedstock(s), the product(s), or both may be individual compounds or mixtures. It is often not worthwhile separating the components in these mixtures completely; specific levels of purity depend on product requirements and process economics.

Operations

Chemical processes may be run in continuous or batch operation.

Batch Operation

In batch operation, production occurs in time-sequential steps in discrete batches. A batch of feedstock(s) is fed (or charged) into a process or unit, then the chemical process takes place, then the product(s) and any other outputs are removed. Such batch production may be repeated over again and again with new batches of feedstock. Batch operation is commonly used in smaller scale plants such as pharmaceutical or specialty chemicals production, for purposes of improved traceability as well as flexibility. Continuous plants are usually used to manufacture commodity or petrochemicals while batch plants are more common in speciality and fine chemical production as well as pharmaceutical active ingredient (API) manufacture.

Continuous Operation

In continuous operation, all steps are ongoing continuously in time. During usual continuous operation, the feeding and product removal are ongoing streams of moving material, which together with the process itself, all take place simultaneously and continuously. Chemical plants or units in continuous operation are usually in a steady state or approximate steady state. Steady state means that quantities related to the process do not change as time passes during operation. Such constant quantities include stream flow rates, heating or cooling rates, temperatures, pressures, and chemical compositions at any given point (location). Continuous operation is more efficient in many large scale operations like petroleum refineries. It is possible for some units to operate continuously and others be in batch operation in a chemical plant. The amount of primary feedstock or product per unit of time which a plant or unit can process is referred to as the capacity of that plant or unit. For examples: the capacity of an oil refinery may be given in terms of barrels of crude oil refined per day; alternatively chemical plant capacity may be given in tons of product produced per day. In actual daily operation, a plant (or unit) will operate at a percentage of its full capacity. Engineers typically assume 90% operating time for plants which work primarily with fluids, and 80% uptime for plants which primarily work with solids.

Units and Fluid Systems

Specific unit operations are conducted in specific kinds of units. Although some units may operate at ambient temperature or pressure, many units operate at higher or lower temperatures or pressures. Vessels in chemical plants are often cylindrical with rounded ends, a shape which can be suited to hold either high pressure or vacuum. Chemical reactions can convert certain kinds of compounds into other compounds in chemical reactors. Chemical reactors may be packed beds and may have solid heterogeneous catalysts which stay in the reactors as fluids move through, or may be simply be stirred vessels in which reactions occur. Since the surface of solid heterogeneous catalysts may sometimes become "poisoned" from deposits such as coke, regeneration of catalysts may be necessary. Fluidized beds may also be used in some cases to ensure good mixing. There can also be units (or subunits) for mixing (including dissolving), separation, heating, cooling, or some combination of these. For example, chemical reactors often have stirring for mixing and heating or cooling to maintain temperature. When designing plants on a large scale, heat produced or absorbed by chemical reactions must be considered. Some plants may have units with organism cultures for biochemical processes such as fermentation or enzyme production.

Distillation unit in Italy

Separation processes include filtration, settling (sedimentation), extraction or leach-ing, distillation, recrystallization or precipitation (followed by filtration or settling), re-verse osmosis, drying, and adsorption. Heat exchangers are often used for heating or cooling, including boiling or condensation, often in conjunction with other units such as distillation towers. There may also be storage tanks for storing feedstock, interme-diate or final products, or waste. Storage tanks commonly have level indicators to show how full they are. There may be structures holding or supporting sometimes massive units and their associated equipment. There are often stairs, ladders, or other steps for personnel to reach points in the units for sampling, inspection, or maintenance. An area of a plant or facility with numerous storage tanks is sometimes called a *tank farm*, especially at an oil depot.

Fluid systems for carrying liquids and gases include piping and tubing of various diam-eter sizes, various types of valves for controlling or stopping flow, pumps for moving or pressurizing liquid, and compressors for pressurizing or moving gases. Vessels, piping, tubing, and sometimes other equipment at high or very low temperature are commonly covered with insulation for personnel safety and to maintain temperature inside. Fluid systems and units commonly have instrumentation such as temperature and pressure sensors and flow measuring devices at select locations in a plant. Online analyzers for chemical or physical property analysis have become more common. Solvents can some-times be used to dissolve reactants or materials such as solids for extraction or leach-ing, to provide a suitable medium for certain chemical reactions to run, or so they can otherwise be treated as fluids.

Chemical Plant Design

Today, the fundamental aspects of designing chemical plants are done by chemical engineers. Historically, this was not always the case and many chemical plants were constructed in a haphazard way before the discipline of chemical engineering became established. Chemical engineering was first established as a profession in the United Kingdom when the first chemical engineering course was given at the University of Manchester in 1887 by George E. Davis in the form of twelve lectures covering various aspects of industrial chemical practice. As a consequence George E. Davis is regarded as the World's first Chemical Engineer.Today Chemical Engineering is a profession and those Professional Chemical Engineers with experience can gain "Chartered" engineer status through the Institution of Chemical Engineers.

Flow diagram for a typical oil refinery

In plant design, typically less than 1 per cent of ideas for new designs ever become commercialized. During this solution process, typically, cost studies are used as an initial screening to eliminate unprofitable designs. If a process appears profitable, then other factors are considered, such as safety, environmental constraints, controllability, etc. The general goal in plant design, is to construct or synthesize "optimum designs" in the neighborhood of the desired constraints.

Many times chemists research chemical reactions or other chemical principles in a laboratory, commonly on a small scale in a "batch-type" experiment. Chemistry information obtained is then used by chemical engineers, along with expertise of their own, to convert to a chemical process and scale up the batch size or capacity. Commonly, a small chemical plant called a pilot plant is built to provide design and operating information before construction of a large plant. From data and operating experience obtained from the pilot plant, a scaled-up plant can be designed for higher or full capacity. After the fundamental aspects of a plant design are determined, mechanical or electrical engineers may become involved with mechanical or electrical details, respectively. Structural engineers may become involved in the plant design to ensure the structures can support the weight of the units, piping, and other equipment.

The units, streams, and fluid systems of chemical plants or processes can be represented by block flow diagrams which are very simplified diagrams, or process flow diagrams which are somewhat more detailed. The streams and other piping are shown as lines with arrow heads showing usual direction of material flow. In block diagrams, units are often simply shown as blocks. Process flow diagrams may use more detailed symbols and show pumps, compressors, and major valves. Likely values or ranges of material flow rates for the various streams are determined based on desired plant capacity using material balance calculations. Energy balances are also done based on heats of reaction, heat capacities, expected temperatures and pressures at various points to calculate amounts of heating and cooling needed in various places and to size heat exchangers. Chemical plant design can be shown in fuller detail in a piping and instrumentation diagram (P&ID) which shows all piping, tubing, valves, and instrumentation, typically with special symbols. Showing a full plant is often complicated in a P&ID, so often only individual units or specific fluid systems are shown in a single P&ID.

In the plant design, the units are sized for the maximum capacity each may have to handle. Similarly, sizes for pipes, pumps, compressors, and associated equipment are chosen for the flow capacity they have to handle. Utility systems such as electric power and water supply should also be included in the plant design. Additional piping lines for non-routine or alternate operating procedures, such as plant or unit startups and shutdowns, may have to be included. Fluid systems design commonly includes isolation valves around various units or parts of a plant so that a section of a plant could be isolated in case of a problem such as a leak in a unit. If pneumatically or hydraulically actuated valves are used, a system of pressurizing lines to the actuators is needed. Any points where process samples may have to be taken should have sampling lines, valves, and access to them included in the detailed design. If necessary, provisions should be made for reducing high pressure or temperature of a sampling stream, such including a pressure reducing valve or sample cooler.

Units and fluid systems in the plant including all vessels, piping, tubing, valves, pumps, compressors, and other equipment must be rated or designed to be able to withstand the entire range of pressures, temperatures, and other conditions which they could possibly encounter, including any appropriate safety factors. All such units and equipment should also be checked for materials compatibility to ensure they can withstand long-term exposure to the chemicals they will come in contact with. Any closed system in a plant which has a means of pressurizing possibly beyond the rating of its equipment, such as heating, exothermic reactions, or certain pumps or compressors, should have an appropriately sized pressure relief valve included to prevent overpressurization for safety. Frequently all of these parameters (temperatures, pressures, flow, etc.) are exhaustively analyzed in combination through a *Hazop* or *fault tree analysis*, to ensure that the plant has no known risk of serious hazard.

Within any constraints the plant is subject to, design parameters are optimized for good economic performance while ensuring safety and welfare of personnel and the surrounding community. For flexibility, a plant may be designed to operate in a range around some optimal design parameters in case feedstock or economic conditions change and re-optimization is desirable. In more modern times, computer simulations or other computer calculations have been used to help in chemical plant design or optimization.

Plant Operation

Process Control

In process control, information gathered automatically from various sensors or other devices in the plant is used to control various equipment for running the plant, thereby controlling operation of the plant. Instruments receiving such information signals and sending out control signals to perform this function automatically are process *controllers*. Previously, pneumatic controls were sometimes used. Electrical controls are now common. A plant often has a control room with displays of parameters such as key temperatures, pressures, fluid flow rates and levels, operating positions of key valves, pumps and other equipment, etc. In addition, operators in the control room can control various aspects of the plant operation, often including overriding automatic control. Process control with a computer represents more modern technology. Based on possible changing feedstock composition, changing products requirements or economics, or other changes in constraints, operating conditions may be re-optimized to maximize profit.

Workers

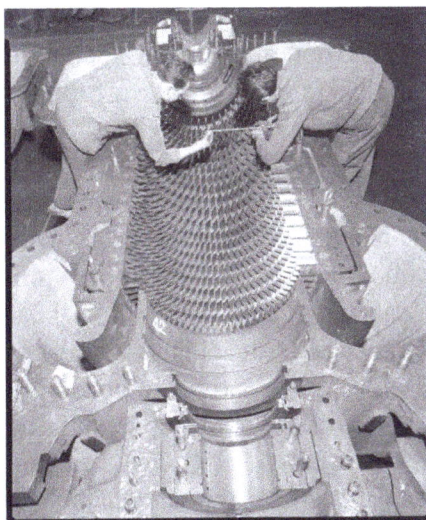

Workers in Italy, 1969. Photo by Paolo Monti

As in any industrial setting, there are a variety of workers working throughout a chemical plant facility, often organized into departments, sections, or other work groups.

Such workers typically include engineers, plant operators, and maintenance technicians. Other personnel at the site could include chemists, management/administration and office workers. Types of engineers involved in operations or maintenance may include chemical process engineers, mechanical engineers for maintaining mechanical equipment, and electrical/computer engineers for electrical or computer equipment.

Transport

Large quantities of fluid feedstock or product may enter or leave a plant by pipeline, railroad tank car, or tanker truck. For example, petroleum commonly comes to a refinery by pipeline. Pipelines can also carry petrochemical feedstock from a refinery to a nearby petrochemical plant. Natural gas is a product which comes all the way from a natural gas processing plant to final consumers by pipeline or tubing. Large quantities of liquid feedstock are typically pumped into process units. Smaller quantities of feedstock or product may be shipped to or from a plant in drums. Use of drums about 55 gallons in capacity is common for packaging industrial quantities of chemicals. Smaller batches of feedstock may be added from drums or other containers to process units by workers.

Maintenance

In addition to feeding and operating the plant, and packaging or preparing the product for shipping, plant workers are needed for taking samples for routine and trouble-shooting analysis and for performing routine and non-routine maintenance. Routine maintenance can include periodic inspections and replacement of worn catalyst, analyzer reagents, various sensors, or mechanical parts. Non-routine maintenance can include investigating problems and then fixing them, such as leaks, failure to meet feed or product specifications, mechanical failures of valves, pumps, compressors, sensors, etc.

Statutory and Regulatory Compliance

When working with chemicals, safety is a concern in order to avoid problems such as chemical accidents . In the United States, the law requires that employers provide workers working with chemicals with access to a Material Safety Data Sheet (MSDS) for every kind of chemical they work with. An MSDS for a certain chemical is prepared and provided by the supplier to whoever buys the chemical. Other laws covering chemical safety, hazardous waste, and pollution must be observed, including statutes such as the Resource Conservation and Recovery Act (RCRA) and the Toxic Substances Control Act (TSCA), and regulations such as the Chemical Facility Anti-Terrorism Standards in the United States. Hazmat (hazardous materials) teams are trained to deal with chemical leaks or spills. Process Hazard Analysis (PHA) is used to assess potential hazards in chemical plants. In 1998, the U. S. Chemical Safety and Hazard Investigation Board has become operational.

Plant Facilities

The actual production or process part of a plant may be indoors, outdoors, or a combination of the two. It may be a traditional stick-built plant or a modular skid. Large modular skids are especially impressive feats of engineering. A modular skid is built including all of the modular equipment needed to do the same job a traditional stick-build plant may perform. However, the modular skid is built within a structural steel frame, allowing it to be shipped to the onsite location without needing to be rebuilt onsite. A modular skid build results in a higher functioning end product, as less hands are required in the onsite setup of the modular skid process unit, resulting in minimized risk for mishaps. The actual production section of a facility usually has the appearance of a rather industrial environment. Hard hats and work shoes are commonly worn. Floors and stairs are often made of metal grating, and there is practically no decoration. There may also be pollution control or waste treatment facilities or equipment. Sometimes existing plants may be expanded or modified based on changing economics, feedstock, or product needs. As in other production facilities, there may be shipping and receiving, and storage facilities. In addition, there are usually certain other facilities, typically indoors, to support production at the site.

Although some simple sample analysis may be able to be done by operations technicians in the plant area, a chemical plant typically has a laboratory where chemists analyze samples taken from the plant. Such analysis can include chemical analysis or determination of physical properties. Sample analysis can include routine quality control on feedstock coming into the plant, intermediate and final products to ensure quality specifications are met. Non-routine samples may be taken and analyzed for investigating plant process problems also. A larger chemical company often has a research laboratory for developing and testing products and processes where there may be pilot plants, but such a laboratory may be located at a site separate from the production plants.

A plant may also have a workshop or maintenance facility for repairs or keeping maintenance equipment. There is also typically some office space for engineers, management or administration, and perhaps for receiving visitors. The decorum there is commonly more typical of an office environment.

Clustering of Commodity Chemical Plants

Chemical Plants used particularly for commodity chemical and petrochemical manufacture,are located in relatively few manufacturing locations around the world largely due to infrastructural needs.This is less important for speciality or fine chemical batch plants. Not all commodity/petrochemicals are produced in any one location but groups of related materials often are, to induce industrial symbiosis as well as material, energy and utility efficiency and other economies of scale. These manufacturing locations often have business clusters of units called chemical plants that share utilities and large scale infrastructure such as power stations, port facilities, road and rail terminals. In the United Kingdom for example there are four main locations for commodity chemical manufacture: near the River Mersey in Northwest England, on the Humber on the East coast of Yorkshire, in Grangemouth near the Firth of Forth in Scotland and on Teesside as part of the Northeast of England Process Industry Cluster (NEPIC). Approximately 50% of the UK's petrochemicals, which are also commodity chemicals, are produced by the industry cluster companies on Teesside at the mouth of the River Tees on three large chemical parks at Wilton, Billingham and Seal Sands.

Corrosion and use of New Materials

Corrosion in chemical process plants is a major issue that consumes billions of dollars yearly. Electrochemical corrosion of metals is pronounced in chemical process plants due to the presence of acid fumes and other electrolytic interactions. Recently, FRP (Fibre-reinforced plastic) is used as a material of construction. The British standard specification BS4994 is widely used for design and construction of the vessels, tanks, etc.

Chemical Thermodynamics

Chemical thermodynamics is the study of the interrelation of heat and work with chemical reactions or with physical changes of state within the confines of the laws of thermodynamics. Chemical thermodynamics involves not only laboratory measurements of various thermodynamic properties, but also the application of mathematical methods to the study of chemical questions and the *spontaneity* of processes.

The structure of chemical thermodynamics is based on the first two laws of thermodynamics. Starting from the first and second laws of thermodynamics, four equations called the "fundamental equations of Gibbs" can be derived. From these four, a multitude of equations, relating the thermodynamic properties of the thermodynamic system can be derived using relatively simple mathematics. This outlines the mathematical framework of chemical thermodynamics.

History

In 1865, the German physicist Rudolf Clausius, in his *Mechanical Theory of Heat*, suggested that the principles of thermochemistry, e.g. the heat evolved in combustion reactions, could be applied to the principles of thermodynamics. Building on the work of Clausius, between the years 1873-76 the American mathematical physicist Willard Gibbs published a series of three papers, the most famous one being the paper *On the Equilibrium of Heterogeneous Substances*. In these papers, Gibbs showed how the first two laws of thermodynamics could be measured graphically and mathematically to determine both the thermodynamic equilibrium of chemical reactions as well as their tendencies to occur or proceed. Gibbs' collection of papers provided the first unified body of thermodynamic theorems from the principles developed by others, such as Clausius and Sadi Carnot.

J. Willard Gibbs - founder of *chemical thermodynamics*

During the early 20th century, two major publications successfully applied the principles developed by Gibbs to chemical processes, and thus established the foundation of the science of chemical thermodynamics. The first was the 1923 textbook *Thermodynamics and the Free Energy of Chemical Substances* by Gilbert N. Lewis and Merle Randall. This book was responsible for supplanting the chemical affinity with the term free energy in the English-speaking world. The second was the 1933 book *Modern Thermodynamics by the methods of Willard Gibbs* written by E. A. Guggenheim. In this manner, Lewis, Randall, and Guggenheim are considered as the founders of modern chemical thermodynamics because of the major contribution of these two books in unifying the application of thermodynamics to chemistry.

Overview

The primary objective of chemical thermodynamics is the establishment of a criterion for the determination of the feasibility or spontaneity of a given transformation. In this manner, chemical thermodynamics is typically used to predict the energy exchanges that occur in the following processes:

1. Chemical reactions

2. Phase changes

3. The formation of solutions

The following state functions are of primary concern in chemical thermodynamics:

- Internal energy (U)

- Enthalpy (H)

- Entropy (S)

- Gibbs free energy (G)

Most identities in chemical thermodynamics arise from application of the first and second laws of thermodynamics, particularly the law of conservation of energy, to these state functions.

The 3 Laws of Thermodynamics:

1. The energy of the universe is constant.

2. In any spontaneous process, there is always an increase in entropy of the universe

3. The entropy of a perfect crystal(well ordered) at 0 Kelvin is zero

Chemical Energy

Chemical energy is the potential of a chemical substance to undergo a transformation through a chemical reaction or to transform other chemical substances. Breaking or making of chemical bonds involves energy or heat, which may be either absorbed or evolved from a chemical system.

Energy that can be released (or absorbed) because of a reaction between a set of chemical substances is equal to the difference between the energy content of the products and the reactants. This change in energy is called the change in internal energy of a chemical reaction. Where $\Delta U^{\circ}_{f\,\text{reactants}}$ is the internal energy of formation of the reactant molecules that can be calculated from the bond energies of the various chemical bonds of the molecules under consideration and $\Delta U^{\circ}_{f\,\text{products}}$ is the internal energy of formation of the product molecules. The change in internal energy is a process which is equal to the heat change if it is measured under conditions of constant volume (at STP condition), as in a closed rigid container such as a bomb calorimeter. However, under conditions of constant pressure, as in reactions in vessels open to the atmosphere, the measured heat change is not always equal to the internal energy change, because pressure-volume work also releases or absorbs energy. (The heat change at constant pressure is called the enthalpy change; in this case the enthalpy of formation).

Another useful term is the heat of combustion, which is the energy released due to a combustion reaction and often applied in the study of fuels. Food is similar to hydrocarbon fuel and carbohydrate fuels, and when it is oxidized, its caloric content is similar In chemical thermodynamics the term used for the chemical potential energy is chemical potential, and for chemical transformation an equation most often used is the Gibbs-Duhem equation.

Chemical Reactions

In most cases of interest in chemical thermodynamics there are internal degrees of freedom and processes, such as chemical reactions and phase transitions, which always create entropy unless they are at equilibrium, or are maintained at a "running equilibrium" through "quasi-static" changes by being coupled to constraining devices, such as pistons or electrodes, to deliver and receive external work. Even for homogeneous "bulk" materials, the free energy functions depend on the composition, as do all the extensive thermodynamic potentials, including the internal energy. If the quantities $\{N_i\}$, the number of chemical species, are omitted from the formulae, it is impossible to describe compositional changes.

Gibbs Function or Gibbs Energy

For a "bulk" (unstructured) system they are the last remaining extensive variables. For an unstructured, homogeneous "bulk" system, there are still various *extensive* compositional variables $\{N_i\}$ that G depends on, which specify the composition, the amounts of each chemical substance, expressed as the numbers of molecules present or (dividing by Avogadro's number = 6.023×10^{23}), the numbers of moles

$$G = G(T, P, \{N_i\}).$$

For the case where only PV work is possible

$$dG = -SdT + VdP + \sum_i \mu_i dN_i$$

in which μ_i is the chemical potential for the i-th component in the system

$$\mu_i = \left(\frac{\partial G}{\partial N_i} \right)_{T,P,N_{j \neq i}, etc.}.$$

The expression for dG is especially useful at constant T and P, conditions which are easy to achieve experimentally and which approximates the condition in living creatures

$$(dG)_{T,P} = \sum_i \mu_i dN_i.$$

Chemical Affinity

While this formulation is mathematically defensible, it is not particularly transparent since one does not simply add or remove molecules from a system. There is always a *process* involved in changing the composition; e.g., a chemical reaction (or many), or movement of molecules from one phase (liquid) to another (gas or solid). We should find a notation which does not seem to imply that the amounts of the components (N_i) can be changed independently. All real processes obey conservation of mass, and in addition, conservation of the numbers of atoms of each kind. Whatever molecules are transferred to or from should be considered part of the "system".

Consequently, we introduce an explicit variable to represent the degree of advancement of a process, a progress variable ξ for the *extent of reaction* (Prigogine & Defay, p. 18; Prigogine, pp. 4–7; Guggenheim, p. 37.62), and to the use of the partial derivative $\partial G/\partial \xi$ (in place of the widely used "ΔG", since the quantity at issue is not a finite change). The result is an understandable expression for the dependence of dG on chemical reactions (or other processes). If there is just one reaction

$$(dG)_{T,P} = \left(\frac{\partial G}{\partial \xi} \right)_{T,P} d\xi.$$

If we introduce the *stoichiometric coefficient* for the *i-th* component in the reaction

$$v_i = \partial N_i / \partial \xi$$

which tells how many molecules of i are produced or consumed, we obtain an algebraic expression for the partial derivative

$$\left(\frac{\partial G}{\partial \xi} \right)_{T,P} = \sum_i \mu_i v_i = -\mathbb{A}$$

where, (De Donder; Progoine & Defay, p. 69; Guggenheim, pp. 37,240), we introduce a concise and historical name for this quantity, the "affinity", symbolized by A, as introduced by Théophile de Donder in 1923. The minus sign comes from the fact the affinity was defined to represent the rule that spontaneous changes will ensue only when the change in the Gibbs free energy of the process is negative, meaning that the chemical species have a positive affinity for each other. The differential for G takes on a simple form which displays its dependence on compositional change

$$(dG)_{T,P} = -\mathbb{A}d\xi.$$

If there are a number of chemical reactions going on simultaneously, as is usually the case

$$(dG)_{T,P} = -\sum_k \mathbb{A}_k d\xi_k.$$

a set of reaction coordinates $\{ \xi_j \}$, avoiding the notion that the amounts of the components (N_i) can be changed independently. The expressions above are equal to zero at thermodynamic equilibrium, while in the general case for real systems, they are negative because all chemical reactions proceeding at a finite rate produce entropy. This can be made even more explicit by introducing the reaction *rates* $d\xi_j/dt$. For each and every *physically independent process* (Prigogine & Defay, p. 38; Prigogine, p. 24)

$$\mathbb{A}\,\dot\xi \le 0.$$

This is a remarkable result since the chemical potentials are intensive system variables, depending only on the local molecular milieu. They cannot "know" whether the temperature and pressure (or any other system variables) are going to be held constant over time. It is a purely local criterion and must hold regardless of any such constraints. Of course, it could have been obtained by taking partial derivatives of any of the other fundamental state functions, but nonetheless is a general criterion for ($-T$ times) the entropy production from that spontaneous process; or at least any part of it that is not captured as external work.

We now relax the requirement of a homogeneous "bulk" system by letting the chemical potentials and the affinity apply to any locality in which a chemical reaction (or any other process) is occurring. By accounting for the entropy production due to irreversible processes, the inequality for dG is now replaced by an equality

$$dG = -SdT + VdP - \sum_k \mathbb{A}_k\, d\xi_k + W'$$

or

$$dG_{T,P} = -\sum_k \mathbb{A}_k\, d\xi_k + W'.$$

Any decrease in the Gibbs function of a system is the upper limit for any isothermal, isobaric work that can be captured in the surroundings, or it may simply be dissipated, appearing as T times a corresponding increase in the entropy of the system and/or its surrounding. Or it may go partly toward doing external work and partly toward creating entropy. The important point is that the *extent of reaction* for a chemical reaction may be coupled to the displacement of some external mechanical or electrical quantity in such a way that one can advance only if the other one also does. The coupling may occasionally be *rigid*, but it is often flexible and variable.

Solutions

In solution chemistry and biochemistry, the Gibbs free energy decrease ($\partial G/\partial\xi$, in molar units, denoted cryptically by ΔG) is commonly used as a surrogate for ($-T$ times) the entropy produced by spontaneous chemical reactions in situations where there is no work being done; or at least no "useful" work; i.e., other than perhaps some $\pm PdV$. The assertion that all *spontaneous reactions have a negative ΔG* is merely a restatement of the fundamental thermodynamic relation, giving it the physical dimensions of energy and somewhat obscuring its significance in terms of entropy. When there is no useful work being done, it would be less misleading to use the Legendre transforms of the entropy appropriate for constant T, or for constant T and P, the Massieu functions $-F/T$ and $-G/T$ respectively.

Non Equilibrium

Generally the systems treated with the conventional chemical thermodynamics are either at equilibrium or near equilibrium. Ilya Prigogine developed the thermodynamic treatment of open systems that are far from equilibrium. In doing so he has discovered phenomena and structures of completely new and completely unexpected types. His generalized, nonlinear and irreversible thermodynamics has found surprising applications in a wide variety of fields.

The non equilibrium thermodynamics has been applied for explaining how ordered structures e.g. the biological systems, can develop from disorder. Even if Onsager's relations are utilized, the classical principles of equilibrium in thermodynamics still show that linear systems close to equilibrium always develop into states of disorder which are stable to perturbations and cannot explain the occurrence of ordered structures.

Prigogine called these systems dissipative systems, because they are formed and maintained by the dissipative processes which take place because of the exchange of energy between the system and its environment and because they disappear if that exchange ceases. They may be said to live in symbiosis with their environment.

The method which Prigogine used to study the stability of the dissipative structures to perturbations is of very great general interest. It makes it possible to study the most varied problems, such as city traffic problems, the stability of insect communities, the development of ordered biological structures and the growth of cancer cells to mention but a few examples.

System Constraints

In this regard, it is crucial to understand the role of walls and other *constraints*, and the distinction between *independent* processes and *coupling*. Contrary to the clear implications of many reference sources, the previous analysis is not restricted to homogeneous, isotropic bulk systems which can deliver only PdV work to the outside world, but applies even to the most structured systems. There are complex systems with many chemical "reactions" going on at the same time, some of which are really only parts of the same, overall process. An *independent* process is one that *could* proceed even if all others were unaccountably stopped in their tracks. Understanding this is perhaps a "thought experiment" in chemical kinetics, but actual examples exist.

A gas reaction which results in an increase in the number of molecules will lead to an increase in volume at constant external pressure. If it occurs inside a cylinder closed with a piston, the equilibrated reaction can proceed only by doing work against an external force on the piston. The extent variable for the reaction can increase only if the piston moves, and conversely, if the piston is pushed inward, the reaction is driven backwards.

Similarly, a redox reaction might occur in an electrochemical cell with the passage of current in wires connecting the electrodes. The half-cell reactions at the electrodes are constrained if no current is allowed to flow. The current might be dissipated as joule heating, or it might in turn run an electrical device like a motor doing mechanical work. An automobile lead-acid battery can be recharged, driving the chemical reaction backwards. In this case as well, the reaction is not an independent process. Some, perhaps most, of the Gibbs free energy of reaction may be delivered as external work.

The hydrolysis of ATP to ADP and phosphate can drive the force times distance work delivered by living muscles, and synthesis of ATP is in turn driven by a redox chain in mitochondria and chloroplasts, which involves the transport of ions across the membranes of these cellular organelles. The coupling of processes here, and in the previous examples, is often not complete. Gas can leak slowly past a piston, just as it can slowly leak out of a rubber balloon. Some reaction may occur in a battery even if no external current is flowing. There is usually a coupling coefficient, which may depend on relative rates, which determines what percentage of the driving free energy is turned into external work, or captured as "chemical work"; a misnomer for the free energy of another chemical process.

Process Simulation

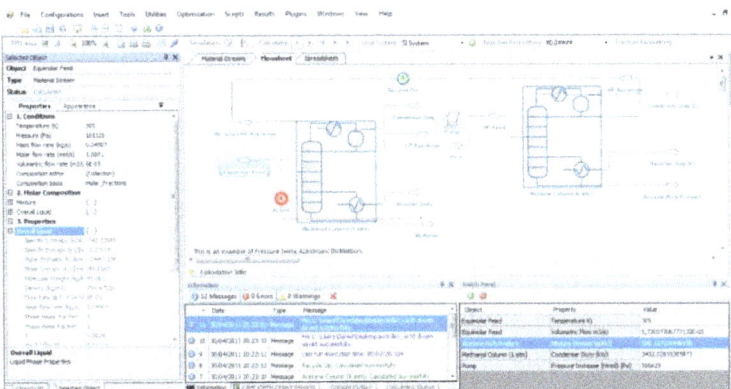

Screenshot of a process simulation software (DWSIM).

Process simulation is used for the design, development, analysis, and optimization of technical processes such as: chemical plants, chemical processes, environmental systems, power stations, complex manufacturing operations, biological processes, and similar technical functions.

Main Principle

Process flow diagram of a typical amine treating process used in industrial plants

Process simulation is a model-based representation of chemical, physical, biological, and other technical processes and unit operations in software. Basic prerequisites are a thorough knowledge of chemical and physical properties of pure components and mixtures, of reactions, and of mathematical models which, in combination, allow the calculation of a process in computers.

Process simulation software describes processes in flow diagrams where unit operations are positioned and connected by product or educt streams. The software has to solve the mass and energy balance to find a stable operating point. The goal of a process simulation is to find optimal conditions for an examined process. This is essentially an optimization problem which has to be solved in an iterative process.

Process simulation always use models which introduce approximations and assumptions but allow the description of a property over a wide range of temperatures and pressures which might not be covered by real data. Models also allow interpolation and extrapolation - within certain limits - and enable the search for conditions outside the range of known properties.

Modelling

The development of models for a better representation of real processes is the core of the further development of the simulation software. Model development is done on the chemical engineering side but also in control engineering and for the improvement of mathematical simulation techniques. Process simulation is therefore one of the few fields where scientists from chemistry, physics, computer science, mathematics, and several engineering fields work together.

VLE of the mixture of Chloroform and Methanol plus NRTL fit and extrapolation to different pressures

A lot of efforts are made to develop new and improved models for the calculation of properties. This includes for example the description of

- thermophysical properties like vapor pressures, viscosities, caloric data, etc. of pure components and mixtures

- properties of different apparatuses like reactors, distillation columns, pumps, etc.

- chemical reactions and kinetics

- environmental and safety-related data

Two main different types of models can be distinguished:

1. Rather simple equations and correlations where parameters are fitted to experimental data.

2. Predictive methods where properties are estimated.

The equations and correlations are normally preferred because they describe the property (almost) exactly. To obtain reliable parameters it is necessary to have experimental data which are usually obtained from factual data banks or, if no data are publicly available, from measurements.

Using predictive methods is much cheaper than experimental work and also than data from data banks. Despite this big advantage predicted properties are normally only used in early steps of the process development to find first approximate solutions and to exclude wrong pathways because these estimation methods normally introduce higher errors than correlations obtained from real data.

Process simulation also encouraged the further development of mathematical models in the fields of numerics and the solving of complex problems.

History

The history of process simulation is strongly related to the development of the computer science and of computer hardware and programming languages. Early working simple implementations of partial aspects of chemical processes were introduced in the 1970s when suitable hardware and software (here mainly the programming languages FORTRAN and C) became available. The modelling of chemical properties began much earlier, notably the cubic equation of states and the Antoine equation were precursory developments of the 19th century.

Steady State and Dynamic Process Simulation

Initially process simulation was used to simulate steady state processes. Steady-state models perform a mass and energy balance of a stationary process (a process in an equilibrium state) it does not depend on time.

Dynamic simulation is an extension of steady-state process simulation whereby time-dependence is built into the models via derivative terms i.e. accumulation of mass and energy. The advent of dynamic simulation means that the time-dependent description, prediction and control of real processes in real time has become possible. This includes the description of starting up and shutting down a plant, changes of conditions during a reaction, holdups, thermal changes and more.

Dynamic simulations require increased calculation time and are mathematically more complex than a steady state simulation. It can be seen as a multiply repeated steady state simulation (based on a fixed time step) with constantly changing parameters.

Dynamic simulation can be used in both an online and offline fashion. The online case being model predictive control, where the real-time simulation results are used to predict the changes that would occur for a control input change, and the control parameters are optimised based on the results. Offline process simulation can be used in the design, troubleshooting and optimisation of process plant as well as the conduction of case studies to assess the impacts of process modifications. Dynamic simulation is also used for operator training.

Chemical Kinetics

Low concentration = Few collisions   High concentration = More collisions

Reaction rate tends to increase with concentration – a phenomenon explained by collision theory.

Chemical kinetics, also known as reaction kinetics, is the study of rates of chemical processes. Chemical kinetics includes investigations of how different experimental conditions can influence the speed of a chemical reaction and yield information about the reaction's mechanism and transition states, as well as the construction of mathematical models that can describe the characteristics of a chemical reaction.

History

In 1864, Peter Waage and Cato Guldberg pioneered the development of chemical kinetics by formulating the law of mass action, which states that the speed of a chemical reaction is proportional to the quantity of the reacting substances.

Van 't Hoff studied chemical dynamics and published in 1884 his famous "Etudes de dynamique chimique". In 1901 he was awarded by the first Nobel Prize in Chemistry "in recognition of the extraordinary services he has rendered by the discovery of the laws of chemical dynamics and osmotic pressure in solutions". After van 't Hoff, chemical kinetics deals with the experimental determination of reaction rates from which rate laws and rate constants are derived. Relatively simple rate laws exist for zero order reactions (for which reaction rates are independent of concentration), first order reactions, and second order reactions, and can be derived for others. Elementary reactions follow the law of mass action, but the rate law of stepwise reactions has to be derived by combining the rate laws of the various elementary steps, and can become rather complex. In consecutive reactions, the rate-determining step often determines the kinetics. In consecutive first order reactions, a steady state approximation can simplify the rate law. The activation energy for a reaction is experimentally determined through the Arrhenius equation and the Eyring equation. The main factors that influence the reaction rate include: the physical state of the reactants, the concentrations of the reactants, the temperature at which the reaction occurs, and whether or not any catalysts are present in the reaction.

Gorban and Yablonsky have suggested that the history of chemical dynamics can be divided into three eras. The first is the van 't Hoff wave searching for the general laws of chemical reactions and relating kinetics to thermodynamics. The second may be called the Semenov--Hinshelwood wave with emphasis on reaction mechanisms, especially for chain reactions. The third is associated with Aris and the detailed mathematical description of chemical reaction networks.

Factors Affecting Reaction Rate

Nature of the Reactants

Depending upon what substances are reacting, the reaction rate varies. Acid/base reactions, the formation of salts, and ion exchange are fast reactions. When covalent bond formation takes place between the molecules and when large molecules are formed, the reactions tend to be very slow. Nature and strength of bonds in reactant molecules greatly influence the rate of its transformation into products.

Physical State

The physical state (solid, liquid, or gas) of a reactant is also an important factor of the rate of change. When reactants are in the same phase, as in aqueous solution, thermal motion brings them into contact. However, when they are in different phases, the reaction is limited to the interface between the reactants. Reaction can occur only at their area of contact; in the case of a liquid and a gas, at the surface of the liquid. Vigorous shaking and stirring may be needed to bring the reaction to completion. This means that the more finely divided a solid or liquid reactant the greater its surface area per unit volume and the more contact it with the other reactant, thus the faster the reaction. To make an analogy, for example, when one starts a fire, one uses wood chips and small branches — one does not start with large logs right away. In organic chemistry, on water reactions are the exception to the rule that homogeneous reactions take place faster than heterogeneous reactions.

Surface Area of Solids

In a solid, only those particles that are at the surface can be involved in a reaction. Crushing a solid into smaller parts means that more particles are present at the surface, and the frequency of collisions between these and reactant particles increases, and so reaction occurs more rapidly. For example, Sherbet (powder) is a mixture of very fine powder of malic acid (a weak organic acid) and sodium hydrogen carbonate. On contact with the saliva in the mouth, these chemicals quickly dissolve and react, releasing carbon dioxide and providing for the fizzy sensation. Also, fireworks manufacturers modify the surface area of solid reactants to control the rate at which the fuels in fireworks are oxidised, using this to create different effects. For example, finely divided aluminium confined in a shell explodes violently. If larger pieces of aluminium are used, the reaction is slower and sparks are seen as pieces of burning metal are ejected.

Concentration

The reactions are due to collisions of reactant species. The frequency with which the molecules or ions collide depends upon their concentrations. The more crowded the molecules are, the more likely they are to collide and react with one another. Thus, an increase in the concentrations of the reactants will usually result in the corresponding increase in the reaction rate, while a decrease in the concentrations will usually have a reverse effect. For example, combustion will occur more rapidly in pure oxygen than in air (21% oxygen).

The rate equation shows the detailed dependence of the reaction rate on the concentrations of reactants and other species present. Different mathematical forms are possible depending on the reaction mechanism. The actual rate equation for a given reaction is determined experimentally and provides information about the reaction mechanism.

Temperature

Temperature usually has a major effect on the rate of a chemical reaction. Molecules at a higher temperature have more thermal energy. Although collision frequency is greater at higher temperatures, this alone contributes only a very small proportion to the increase in rate of reaction. Much more important is the fact that the proportion of reactant molecules with sufficient energy to react (energy greater than activation energy: $E > E_a$) is significantly higher and is explained in detail by the Maxwell–Boltzmann distribution of molecular energies.

The 'rule of thumb' that the rate of chemical reactions doubles for every 10 °C temperature rise is a common misconception. This may have been generalized from the special case of biological systems, where the α (temperature coefficient) is often between 1.5 and 2.5.

A reaction's kinetics can also be studied with a temperature jump approach. This involves using a sharp rise in temperature and observing the relaxation time of the return to equilibrium. A particularly useful form of temperature jump apparatus is a shock tube, which can rapidly jump a gas's temperature by more than 1000 degrees.

Catalysts

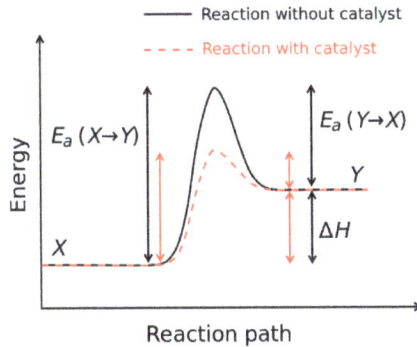

Generic potential energy diagram showing the effect of a catalyst in a hypothetical endothermic chemical reaction. The presence of the catalyst opens a different reaction pathway (shown in red) with a lower activation energy. The final result and the overall thermodynamics are the same.

A catalyst is a substance that alters the rate of a chemical reaction but remains chemically unchanged afterwards. The catalyst increases the rate of the reaction by providing a different reaction mechanism to occur with a lower activation energy. In autocatalysis a reaction product is itself a catalyst for that reaction leading to positive feedback. Proteins that act as catalysts in biochemical reactions are called enzymes. Michaelis–Menten kinetics describe the rate of enzyme mediated reactions. A catalyst does not affect the position of the equilibrium, as the catalyst speeds up the backward and forward reactions equally.

In certain organic molecules, specific substituents can have an influence on reaction rate in neighbouring group participation.

Pressure

Increasing the pressure in a gaseous reaction will increase the number of collisions between reactants, increasing the rate of reaction. This is because the activity of a gas is directly proportional to the partial pressure of the gas. This is similar to the effect of increasing the concentration of a solution.

In addition to this straightforward mass-action effect, the rate coefficients themselves can change due to pressure. The rate coefficients and products of many high-temperature gas-phase reactions change if an inert gas is added to the mixture; variations on this effect are called fall-off and chemical activation. These phenomena are due to exothermic or endothermic reactions occurring faster than heat transfer, causing the reacting molecules to have non-thermal energy distributions (non-Boltzmann distribution). Increasing the pressure increases the heat transfer rate between the reacting molecules and the rest of the system, reducing this effect.

Condensed-phase rate coefficients can also be affected by (very high) pressure; this is a completely different effect than fall-off or chemical-activation. It is often studied using diamond anvils.

A reaction's kinetics can also be studied with a pressure jump approach. This involves making fast changes in pressure and observing the relaxation time of the return to equilibrium.

Experimental Methods

The experimental determination of reaction rates involves measuring how the concentrations of reactants or products change over time. For example, the concentration of a reactant can be measured by spectrophotometry at a wavelength where no other reactant or product in the system absorbs light.

For reactions which take at least several minutes, it is possible to start the observations after the reactants have been mixed at the temperature of interest.

Fast Reactions

For faster reactions, the time required to mix the reactants and bring them to a specified temperature may be comparable or longer than the half-life of the reaction. Special methods to start fast reactions without slow mixing step include

- Stopped flow methods, which can reduce the mixing time to the order of a millisecond

- Chemical relaxation methods such as temperature jump and pressure jump, in which a pre-mixed system initially at equilibrium is perturbed by rapid heating or depressurization so that it is no longer at equilibrium, and the relaxation back to equilibrium is observed. For example, this method has been used to study the neutralization $H_3O^+ + OH^-$ with a half-life of 1 µs or less under ordinary conditions.

- Flash photolysis, in which a laser pulse produces highly excited species such as free radicals, whose reactions are then studied.

Equilibrium

While chemical kinetics is concerned with the rate of a chemical reaction, thermodynamics determines the extent to which reactions occur. In a reversible reaction, chemical equilibrium is reached when the rates of the forward and reverse reactions are equal (the principle of detailed balance) and the concentrations of the reactants and products no longer change. This is demonstrated by, for example, the Haber–Bosch process for combining nitrogen and hydrogen to produce ammonia. Chemical clock reactions such as the Belousov–Zhabotinsky reaction demonstrate that component concentrations can oscillate for a long time before finally attaining the equilibrium.

Free Energy

In general terms, the free energy change (ΔG) of a reaction determines whether a chemical change will take place, but kinetics describes how fast the reaction is. A reaction can be very exothermic and have a very positive entropy change but will not happen in practice if the reaction is too slow. If a reactant can produce two different products, the thermodynamically most stable one will in general form, except in special circumstances when the reaction is said to be under kinetic reaction control. The Curtin–Hammett principle applies when determining the product ratio for two reactants interconverting rapidly, each going to a different product. It is possible to make predictions about reaction rate constants for a reaction from free-energy relationships.

The kinetic isotope effect is the difference in the rate of a chemical reaction when an atom in one of the reactants is replaced by one of its isotopes.

Chemical kinetics provides information on residence time and heat transfer in a chemical reactor in chemical engineering and the molar mass distribution in polymer chemistry.

Applications and Models

The mathematical models that describe chemical reaction kinetics provide chemists and chemical engineers with tools to better understand and describe chemical processes such as food decomposition, microorganism growth, stratospheric ozone decomposition, and the complex chemistry of biological systems. These models can also be used in the design or modification of chemical reactors to optimize product yield, more efficiently separate products, and eliminate environmentally harmful by-products. When performing catalytic cracking of heavy hydrocarbons into gasoline and light gas, for example, kinetic models can be used to find the temperature and pressure at which the highest yield of heavy hydrocarbons into gasoline will occur.

Chemical Kinetics is frequently validated and explored through modeling in specialized packages as a function of ordinary differential equation-solving (ODE-solving) and curve-fitting.

Process Flow Diagram

A process flow diagram (PFD) is a diagram commonly used in chemical and process engineering to indicate the general flow of plant processes and equipment. The PFD displays the relationship between *major* equipment of a plant facility and does not show minor details such as piping details and designations. Another commonly used term for a PFD is a *flowsheet*.

Typical Content of a Process Flow Diagram

Some typical elements from process flow diagrams, as provided by the open source program, Dia. Click for image legend.

Typically, process flow diagrams of a single unit process will include the following:

- Process piping
- Major equipment items
- Control valves and other major valves
- Connections with other systems
- Major bypass and recirculation streams
- Operational data (temperature, pressure, mass flow rate, density, etc.), often by stream references to a mass balance.
- Process stream names

Process flow diagrams generally do not include:

- Pipe classes or piping line numbers
- Process control instrumentation (sensors and final elements)
- Minor bypass lines
- Isolation and shutoff valves
- Maintenance vents and drains
- Relief and safety valves
- Flanges

Process flow diagrams of multiple process units within a large industrial plant will usually contain less detail and may be called *block flow diagrams* or *schematic flow diagrams*.

Process Flow Diagram Examples

The process flow diagram below depicts a single chemical engineering unit process known as an amine treating plant:

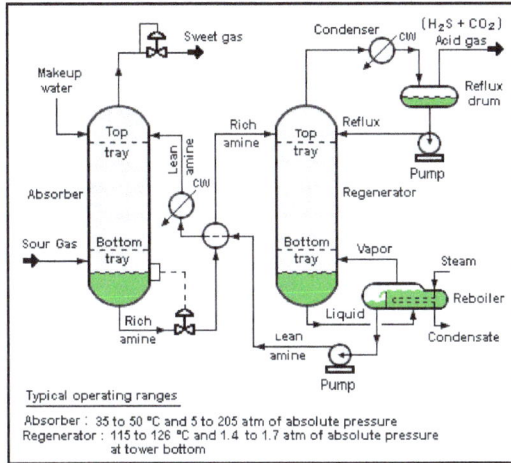

Flow diagram of a typical amine treating process used in industrial plants

Multiple Process Units Within an Industrial Plant

The process flow diagram below is an example of a schematic or block flow diagram and depicts the various unit processes within a typical oil refinery:

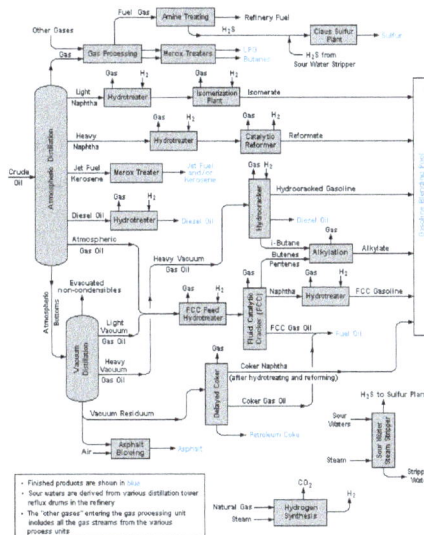

A typical oil refinery-SL

Other Items of Interest

A PFD can be computer generated from process simulators, CAD packages, or flow chart software using a library of chemical engineering symbols. Rules and symbols are available from standardization organizations such as DIN, ISO or ANSI. Often PFDs are produced on large sheets of paper.

PFDs of many commercial processes can be found in the literature, specifically in encyclopedias of chemical technology, although some might be outdated. To find recent ones, patent databases such as those available from the United States Patent and Trademark Office can be useful.

Standards

- ISO 10628: Flow Diagrams For Process Plants - General Rules
- ANSI Y32.11: Graphical Symbols For Process Flow Diagrams (withdrawn 2003)
- SAA AS 1109: Graphical Symbols For Process Flow Diagrams For The Food Industry

Process Miniaturization

Chemical process miniaturization refers to a philosophical concept within the discipline of process design that challenges the notion of "economy of scale" or "bigger is better". In this context, process design refers to the discipline taught primarily to chemical engineers. However, the emerging discipline of process miniaturization will involve integrated knowledge from many areas; as examples, systems engineering and design, remote measurement and control using intelligent sensors, biological process systems engineering, and advanced manufacturing robotics, etc.

One of the challenges of chemical engineering has been to design processes based on chemical laboratory-scale methods, and to scale-up processes so that products can be manufactured that are economically affordable.

As a process becomes larger, more product can be produced per unit time, so when a process technology becomes established or mature, and operates consistently without upsets or "downtime", more economic efficiency can be gained from scale-up. Given a fixed price for the feedstock (e.g. the price per barrel of crude oil), the product cost can be decreased using a larger scale process because the capital investment and operational costs do not normally increase linearly with scale. For example, the capacity or volume of a cylindrical vessel used to produce a product increases proportional to the square of the radius of the cylinder, so cost of materials per unit volume decreases. But the costs to design and fabricate the vessel have traditionally been less sensitive to scale. In other words, one can design a small vessel and fabricate it for about the same cost as the larger vessel. In addition, the cost to control and operate a process (or a process unit component) does not change substantially with the scale. For example, if it takes one operator to operate a small process, that same operator can probably operate the larger process.

The economy of scale concept, as taught to chemical engineers, has led to the notion that one of the objectives of process development and design is to achieve "economy of scale" by scaling-up to the largest possible size processing plant so that the product cost can be economically affordable. This disciplinary philosophy has been reinforced by example designs in the petroleum refining and petrochemical industries, where feed-stocks have been transported as fluids in pipelines, large tanker ships, and railcars.

Fluids, by definition are materials that flow and can be transferred using pumps or gravity. Therefore, large pumps, valves, and pipelines exist to transfer large amounts of fluids in the process industries. Process miniaturization, in contrast, will involve processing of large amounts of solids from renewable biomass resources; therefore, new thinking towards process designs optimized for solids processing will be required.

The concept of a microprocess has been defined by S. S. Sofer while a professor at the New Jersey Institute of Technology. A microprocess has the following characteristics:

1) Portability

2) Capable of being mass produced using advanced robotic manufacturing methods

3) Approaching total automation

4) A new technology

Miniaturization of Electronic Devices

The microprocess design philosophy has been largely envisioned by historical analysis of the role that component miniaturization has played in the information technology industry. It is the evolution of the miniaturization of computer hardware that has enabled the thinking about process miniaturization, in the chemical engineering design context. Rather than the traditional design objective as "scale-up" of processing to one centralized large processing plant (e.g. the mainframe), one can envision achieving the economic objectives using a "scale-out" philosophy (e.g. multiple microcomputers).

Electrical and electronic devices have always played an important role in chemical process plant automation. However, initially, simple thermometers such as those containing mercury, and pressure gauges which were completely mechanical in nature were used to monitor process conditions (such as the temperature, pressure and level in a chemical reactor). Process conditions were adjusted based largely on a human operator's heuristic knowledge of the process behavior. Even with electronic automation installed, many process still require substantial operator interaction, particularly during the start-up phase of the process, or during deployment of a new technology.

Process control of the future will involve the widespread utilization of intelligent sensors, and mass-produced intelligent miniaturized devices such as programmable logic controllers that communicate wirelessly to process actuators. Since these devices will be miniaturized to reduce manufacturing cost, this enables the devices to be embedded in structures so that they become invisible to the casual observer. The cost of such sensors will likely be reduced to a point where they either "function or don't function". When that cost threshold has been reached, the repair procedure will be to disable the sensor, and to actuate a redundant working sensor. In otherwords, entire complex control systems will become so low cost, that repair will not be economically viable.

The intelligence of the process will be developed using process simulation models based on scientific fundamentals. Heuristic rules will be programmed into the micro-controllers, which will largely eliminate the need for constant monitoring by human heuristic knowledge of the process behavior. Process which can automatically self-optimize through advanced algorithms developed by microprocess engineers will be embedded, and only accessible to the knowledge-owner. This will enable the construction of large networks of automous microprocesses.

Process Miniaturization for Knowledge-based Businesses

Advanced process control systems for process miniaturization will increase the need for controlling the security and ownership of process intelligence in a knowledge-based business. It will become more difficult to control intellectual property through the traditional method of patents; therefore, trademarks, brand recognition, and copyright laws will play a more important role in value security for knowledge-based businesses of the future.

Techno-economic analysis, as taught in traditional chemical process design, will also dramatically shift from a conservative viewpoint of utilization of historical trend economics and cash flow analysis. Economic viability of a given enterprise will be more linked to acquisition of real-time economic information, that can rapidly change based on empirical observations created by an emerging discipline of microprocess development systems; therefore, the models will be more based on "what can be?" rather that "what has the past shown?"

Process Miniaturization for Future Societies Based on Renewable Materials

Rather than one large central plant, that has to be fed a large amount of feedstock, such as a refinery that can unload a tanker shipment of petroleum if located next to an ocean, the discipline of process miniaturization envisions the distribution of the process technology to areas where the feedstock is not readily transportable in large quantities to a large centralized processing plant. The miniaturized process technology may simply involve transformation of solid biomass materials from multiple distributed microprocesses into more easily manageable fluids. The fluids can then be transported or distributed to larger-scale intelligent processing nodes using conventional fluid transport technology.

Historically, small processes or microprocesses *per se* have always existed. For example, small vineyards and breweries have produced feedstock, processed it, and stored product in what could be considered "microprocess" when compared to processes designed based on the petrochemical industry model or, for example, large-scale production of beer. Small villages in India and other places in the world have learned to produce biogas from animal manure in what could be considered small-scale microprocesses for the production of energy. However, microprocesses and process miniaturization as a design philosophy includes the notion of approaching total automation, and is a new technology which has been enabled by computer hardware miniaturization, for example, the microprocessor. It is easy to envision processes which can be mass-produced and transported. For example, many appliances such as air conditioners, domestic washing machines, and refrigerators could be considered microprocesses.

The design philosophy of process miniaturization envisions that "scale-down" of complex processes involving multiple process unit operations can be achieved, and that economy of scale will be more related to the size of a network of distributed autonomous microprocesses. Since failure of one autonomous microprocess does not cause shutdown of the entire network, microprocesses will lead to more economically efficient, robust, and stable production of products that have traditionally been produced for a petroleum-based society.

Since fossil fuels by definition are being consumed and are non-renewable, future fuel and materials will be based on renewable biomass.

Process Miniaturization for Microbial Fuel Cells

The conversion of biomass into energy is perhaps more challenging to the technologist than energy from fossil fuels. Water, dissolved organic and inorganic compounds, and solid particulates of various size can be present in biomass processes. It is perhaps the development of microbial fuel cells where the philosophical thinking of process miniaturization will play a wider role. Distribution of knowledge, in a fashionable, intriguing style through miniaturized devices, can be substantially enhanced (accelerated) by low power consuming devices (such as smart phones). A rethinking of "what is a power-plant?" can create enormous innovations, given recent advances in membrane materials of construction, immobilized whole cell methodologies, metabolic engineering, and nanotechnology.

The challenges of microbial fuel cells relate mainly to finding lower cost manufacturing methods, materials of construction, and systems design. Bruce Logan from the Penn State University has described in several research articles and reviews these challenges.

However, even with existing designs which generate low power, there are applications in distribution of electrical recharging systems to remote areas of Africa, where smart phone, can enable access to the vast information of the internet, and to provide lighting. These systems can run on agricultural, animal and human waste streams using naturally occurring bacteria.

Process Miniaturization for Mini Nuclear Reactors

Nuclear power is considered "green technology" in that it does not produce carbon dioxide, a green house gas, as do traditional natural gas or coal-fired power plants. The economics of the deployment of mini nuclear reactors has been discussed in an article in "The Economist".

The advantages of mini nuclear reactors has also been discussed by Secretary of Energy, Steven Chu. As discussed by Chu, the reactors would be manufactured in a factory-like situation and then transported, intact by rail or ship to different parts of the country or world. Economy of scale by size is replaced by economy of scale by number. Many companies are not willing to accept the risk of investing $8B to $9B dollars in single large reactor, so one of the most attractive feactures of process miniaturization is a reduction in the risk of capital investment, and the possibility of recovering investment by reselling and relocating a functional turn-key microprocess to a new owner - a major economic advantage of the portability of microprocesses.

References

- Truskey, George; Yuan F; Katz D. Transport Phenomena in Biological Systems (Second ed.). Prentice Hall. p. 888. ISBN 978-0131569881.

- Plawsky, Joel L. (April 2001). Transport phenomena fundamentals (Chemical Industries Series). CRC Press. pp. 1, 2, 3. ISBN 978-0-8247-0500-8.

- Lienhard, John H. (2011). A Heat Transfer Textbook. Dover Civil and Mechanical Engineering Series. Courier Corporation. ISBN 9780486479316.

- Duong, Do D (1998). "Fundamentals of Diffusion and Adsorption in Porus Media". Adsorption Analysis: Equilibria and Kinetics. Series on Chemical Engineering. 2. World Scientific. ISBN 9781783262243.

- Kirwan, Donald J. (1987). "Mass transfer principles". In Rousseau, Ronald W. Handbook of Separation Process Technology. John Wiley & Sons. ISBN 9780471895589.

- Taylor, Ross; Krishna, R. (1993). Multicomponent Mass Transfer. Wiley Series in Chemical Engineering. 2. John Wiley & Sons. ISBN 9780471574170.

- Ott, Bevan J.; Boerio-Goates, Juliana (2000). Chemical Thermodynamics – Principles and Applications. Academic Press. ISBN 0-12-530990-2.

- Steinfeld J.I., Francisco J.S. and Hase W.L. Chemical Kinetics and Dynamics (2nd ed., Prentice-Hall 1999) p.140-3 ISBN 0-13-737123-3

- Espenson, J.H. Chemical Kinetics and Reaction Mechanisms (2nd ed., McGraw-Hill 2002), p.264-6 ISBN 0-07-288362-6

- A.N. Gorban, G.S. Yablonsky Three Waves of Chemical Dynamics, Mathematical Modelling of Natural Phenomena 10(5) (2015), p. 1–5.

- Delgass; et al. "Seventy Five Years of Chemical Engineering". Prudue University. Retrieved 13 August 2013.

- "Year in Science, 2010", Interview with Secretary of Energy, Steven Chu, Discover magazine, Jan-Feb 2011, p. 42.

Applications of Chemical Engineering

Chemical engineering has diverse applications. Some of these are food engineering, plastics engineering, tissue engineering, ceramic engineering, cheminformatics and process engineering. Plastics engineering deals with the designing and manufacturing of plastic products and food engineering deals includes agricultural engineering, mechanical and chemical engineering. The diverse applications of chemical engineering in the current scenario have been thoroughly discussed in this chapter.

Food Engineering

Bread factory in Germany

Food engineering is a multidisciplinary field of applied physical sciences which combines science, microbiology, and engineering education for food and related industries. Food engineering includes, but is not limited to, the application of agricultural engineering, mechanical engineering and chemical engineering principles to food materials. Food engineers provide the technological knowledge transfer essential to the cost-effective production and commercialization of food products and services. Physics, chemistry, and mathematics are fundamental to understanding and engineering products and operations in the food industry.

Food engineering encompasses a wide range of activities. Food engineers are employed in food processing, food machinery, packaging, ingredient manufacturing, instrumentation, and control. Firms that design and build food processing plants, consulting firms, government agencies, pharmaceutical companies, and health-care firms also employ food engineers. Specific food engineering activities include:

- drug/food products;

- design and installation of food/biological/pharmaceutical production processes;

- design and operation of environmentally responsible waste treatment systems;

- marketing and technical support for manufacturing plants.

Topics in Food Engineering

In the development of food engineering, one of the many challenges is to employ modern tools, technology, and knowledge, such as computational materials science and nanotechnology, to develop new products and processes. Simultaneously, improving quality, safety, and security remain critical issues in food engineering study. New packaging materials and techniques are being developed to provide more protection to foods, and novel preservation technology is emerging. Additionally, process control and automation regularly appear among the top priorities identified in food engineering. Advanced monitoring and control systems are developed to facilitate automation and flexible food manufacturing. Furthermore, energy saving and minimization of environmental problems continue to be important food engineering issues, and significant progress is being made in waste management, efficient utilization of energy, and reduction of effluents and emissions in food production.

Typical topics include:food people

- Advances in classical unit operations in engineering applied to food manufacturing

- Progresses in the transport and storage of liquid and solid foods

- Developments in heating, chilling and freezing of foods

- Advanced mass transfer in foods

- New chemical and biochemical aspects of food engineering and the use of kinetic analysis

- New techniques in dehydration, thermal processing, non-thermal processing, extrusion, liquid food concentration, membrane processes and applications of membranes in food processing

- Shelf-life, electronic indicators in inventory management, and sustainable technologies in food processing

- Modern packaging, cleaning, and sanitation technologies.

- Development of sensors systems for quality and safety assessment

Plastics Engineering

Plastics engineering encompasses the processing, design, development, and manufacture of plastics products. A plastic is a polymeric material that is in a semi-liquid state, having the property of plasticity and exhibiting flow. Plastics engineering encompasses plastics material and plastic machinery. Plastic Machinery is the general term for all types of machinery and devices used in the plastics processing industry. The nature of plastic materials poses unique challenges to an engineer. Mechanical properties of plastics are often difficult to quantify, and the plastics engineer has to design a product that meets certain specifications while keeping costs to a minimum. Other properties that the plastics engineer has to address include: outdoor weatherability, thermal properties such as upper use temperature, electrical properties, barrier properties, and resistance to chemical attack.

In plastics engineering, as in most engineering disciplines, the economics of a product plays an important role. The cost of plastic materials ranges from the cheapest commodity plastics used in mass-produced consumer products to the very expensive, specialty plastics. The cost of a plastic product is measured in different ways, and the absolute cost of a plastic material is difficult to ascertain. Cost is often measured in price per pound of material, or price per unit volume of material. In many cases however, it is important for a product to meet certain specifications, and cost could then be measured in price per unit of a property. Price with respect to processibility is often important, as some materials need to be processed at very high temperatures, increasing the amount of cooling time a part needs. In a large production run cooling time is very expensive.

Some plastics are manufactured from re-cycled materials but their use in engineering tends to be limited because the consistency of formulation and their physical properties tend to be less consistent. Electrical and electronic equipment and motor vehicle markets together accounted for 58 percent of engineered plastics demand in 2003. Engineered plastics demand in the US was estimated at $9,702 million in 2007.

A big challenge for plastics engineers is the reduction of the ecological footprints of their products. First attempts like the Vinyloop process can guarantee that a product's primary energy demand is 46 percent lower than conventional produced PVC. The global warming potential is 39 percent lower.

Plastics Engineering Specialties

- Consumer Plastics

- Medical plastics

- Automotive plastics

- Recycled or recyclable plastics

- Biodegradable plastics

- Elastomers / rubber

- Epoxys

- Plastics processing: injection moulding, plastics extrusion, stretch-blow molding, thermoforming, compression molding, calendering, transfer molding, laminating, fiberglass molding, pultrusion, filament winding, vacuum forming, rotational molding

- Ultrasonic welding

Pharmaceutical Engineering

Pharmaceutical engineering is a branch of pharmaceutical science and technology that involves development and manufacturing of products, processes, and components in the pharmaceuticals industry (i.e. drugs & biologics). While developing pharmaceutical products involves many interrelated disciplines (e.g. medicinal chemists, analytical chemists, clinicians/pharmacologists, pharmacists, chemical engineers, biomedical engineers, etc.), the specific subfield of "pharmaceutical engineering" has only emerged recently as a distinct engineering discipline. This now brings the problem-solving principles and quantitative training of engineering to complement the other scientific fields already involved in drug development.

Academic Programs

There are still relatively few academic programs with this explicit focus. The first one began at the University of Michigan, as a joint project between their College of Engineering and School of Pharmacy. Because such programs are not yet common, many pharmaceutical engineers have had their formal engineering training in chemical or biomedical engineering.

Most pharmaceutical engineering programs are graduate-level, and as with biomedical engineering there is generally an expectation that engineers and scientists working in pharmaceutical engineering should have some relevant graduate-level education. Many have a masters or PhD degree in chemical or biomedical engineering, or a related science. In Italy there is a university degree course (5 years) in Chemistry and Pharmaceutical Technologies (Chimica e Tecnologie Farmaceutiche), different from pharmacy, that ability as a pharmacist and different roles in the industry as an engineer (for the Italian legislation is not really an engineer though performs the same tasks).

Professional Licensure and Certification

In most jurisdictions, engineering licensure (e.g. Licensed "Professional Engineer" or P.E.) is not discipline-specific, so any licensed engineer with competency in pharmaceutical engineering may qualify as licensed. However, in the U.S., most pharmaceutical engineers fall under the "industrial exemption," which does not require a (P.E.) license for those engineers whose work is completely internal and for a private employer. There are ongoing debates about whether to narrow or eliminate this exemption from engineering licensure, and the Executive Director of the National Society of Professional Engineers (NSPE) recently advocated requiring licensure for engineers in the pharmaceutical industry (among a few others).

Most U.S. jurisdictions require two examinations as part of their licensing criteria, the second of which allows electing a particular discipline of emphasis (while not affecting the license itself, as noted above), so pharmaceutical engineers are likely to select the testing option of either chemical engineering or biomedical engineering (which are generally considered the closest available options).

There is also a private (non-governmental) certification offered by the professional organization International Society for Pharmaceutical Engineering (ISPE), known as Certified Pharmaceutical Industry Professional (CPIP). This tends to focus more on (later-stage) manufacturing and commercialization issues, etc., rather than early-stage things like drug design, discovery assays, and preclinical development.

Common Specialties

- Pharmaceutical Development Sciences - broadly, assays or techniques for discovering, modifying, or designing drug substances or excipients; in particular, rational drug design - as a relatively recent alternative to traditional trial-and-error drug discovery processes - relies upon principles of engineering more than many other pharmaceutical sciences such as formulation or medicinal chemistry.

- Bio-/Pharmaceutical Manufacturing Science - optimal processes for producing drug substances & products with quality and efficiency

- Clinical Science - applying engineering principles toward conduct of studies to assess safety & efficacy, for the medical community and regulators

- Regulatory Science - scientific bases for regulatory decision-making (typically by the FDA, in the U.S.), with an emphasis on risk-benefit analysis

- Pharmaceutical Devices - designing instruments, tools, or implants which facilitate the making, handling, or use of drugs (e.g. drug delivery chips)

These specialties overlap with other engineering areas as well as non-engineering scientific and medical fields, although in all specialties Pharmaceutical Engineers tend to have a distinct focus on product and process design and quantitative analysis. And in addition to these technical areas, some pharmaceutical engineers pursue careers as business or legal professionals. Their scientific and engineering background is often suitable for careers in management, patent law, or even entrepreneurship - for example.

Tissue Engineering

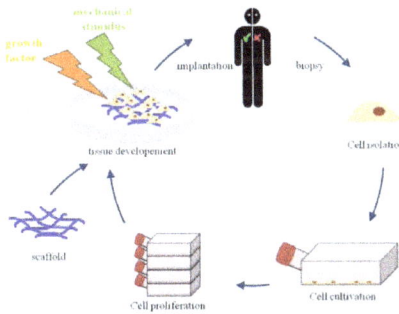

Principle of tissue engineering

Tissue engineering is the use of a combination of cells, engineering and materials methods, and suitable biochemical and physicochemical factors to improve or replace biological tissues. Tissue engineering involves the use of a scaffold for the formation of new viable tissue for a medical purpose. While it was once categorized as a sub-field of biomaterials, having grown in scope and importance it can be considered as a field in its own.

While most definitions of tissue engineering cover a broad range of applications, in practice the term is closely associated with applications that repair or replace portions of or whole tissues (i.e., bone, cartilage, blood vessels, bladder, skin, muscle etc.). Often, the tissues involved require certain mechanical and structural properties for proper functioning. The term has also been applied to efforts to perform specific biochemical functions using cells within an artificially-created support system (e.g. an artificial pancreas, or a bio artificial liver). The term regenerative medicine is often used synonymously with tissue engineering, although those involved in regenerative medicine place more emphasis on the use of stem cells or progenitor cells to produce tissues.

Overview

A commonly applied definition of tissue engineering, as stated by Langer and Vacanti, is "an interdisciplinary field that applies the principles of engineering and life sciences toward the development of biological substitutes that restore, maintain, or improve

[Biological tissue] function or a whole organ". Tissue engineering has also been defined as "understanding the principles of tissue growth, and applying this to produce functional replacement tissue for clinical use." A further description goes on to say that an "underlying supposition of tissue engineering is that the employment of natural biology of the system will allow for greater success in developing therapeutic strategies aimed at the replacement, repair, maintenance, and/or enhancement of tissue function."

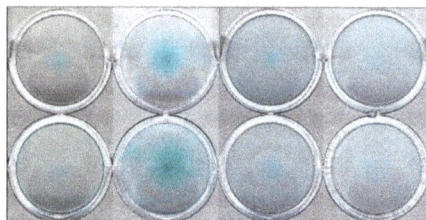

Micro-mass cultures of C3H-10T1/2 cells at varied oxygen tensions stained with Alcian blue

Powerful developments in the multidisciplinary field of tissue engineering have yielded a novel set of tissue replacement parts and implementation strategies. Scientific advances in biomaterials, stem cells, growth and differentiation factors, and biomimetic environments have created unique opportunities to fabricate tissues in the laboratory from combinations of engineered extracellular matrices ("scaffolds"), cells, and biologically active molecules. Among the major challenges now facing tissue engineering is the need for more complex functionality, as well as both functional and biomechanical stability and vascularization in laboratory-grown tissues destined for transplantation. The continued success of tissue engineering, and the eventual development of true human replacement parts, will grow from the convergence of engineering and basic research advances in tissue, matrix, growth factor, stem cell, and developmental biology, as well as materials science and bio informatics.

In 2003, the NSF published a report entitled "The Emergence of Tissue Engineering as a Research Field", which gives a thorough description of the history of this field.

Examples

- Bioartificial windpipe: The first procedure of regenerative medicine of an implantation of a "bioartificial" organ.

- In vitro meat: Edible artificial animal muscle tissue cultured in vitro.

- Bioartificial liver device: several research efforts have produced hepatic assist devices utilizing living hepatocytes.

- Artificial pancreas: research involves using islet cells to produce and regulate insulin, particularly in cases of diabetes.

- Artificial bladders: Anthony Atala (Wake Forest University) has successfully implanted artificially grown bladders into seven out of approximately 20 human test subjects as part of a long-term experiment.

- Cartilage: lab-grown tissue was successfully used to repair knee cartilage.

- Scaffold-free cartilage: Cartilage generated without the use of exogenous scaffold material. In this methodology, all material in the construct is cellular or material produced directly by the cells themselves.

- Doris Taylor's heart in a jar

- Tissue-engineered airway

- Tissue-engineered vessels

- Artificial skin constructed from human skin cells embedded in a hydrogel, such as in the case of bioprinted constructs for battlefield burn repairs.

- Artificial bone marrow

- Artificial bone

- Laboratory-grown penis

- Oral mucosa tissue engineering

- Foreskin

Cells as Building Blocks

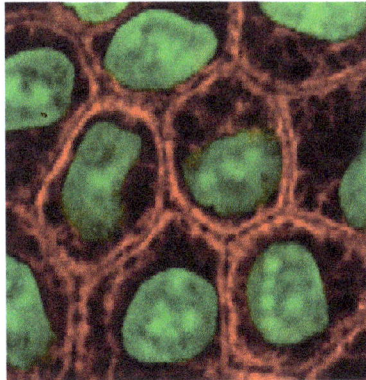

Stained cells in culture

Tissue engineering utilizes living cells as engineering materials. Examples include using living fibroblasts in skin replacement or repair, cartilage repaired with living chondrocytes, or other types of cells used in other ways.

Cells became available as engineering materials when scientists at Geron Corp. discovered how to extend telomeres in 1998, producing immortalized cell lines. Before this, laboratory cultures of healthy, noncancerous mammalian cells would only divide a fixed number of times, up to the Hayflick limit, before dying.

Extraction

From fluid tissues such as blood, cells are extracted by bulk methods, usually centrifugation or apheresis. From solid tissues, extraction is more difficult. Usually the tissue is minced, and then digested with the enzymes trypsin or collagenase to remove the extracellular matrix (ECM) that holds the cells. After that, the cells are free floating, and extracted using centrifugation or apheresis. Digestion with trypsin is very dependent on temperature. Higher temperatures digest the matrix faster, but create more damage. Collagenase is less temperature dependent, and damages fewer cells, but takes longer and is a more expensive reagent.

Types of Cells

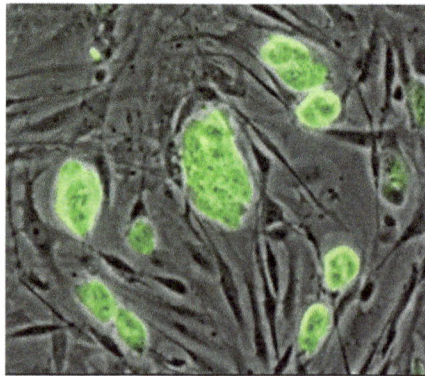

Mouse embryonic stem cells

Cells are often categorized by their source:

- Autologous cells are obtained from the same individual to which they will be re-implanted. Autologous cells have the fewest problems with rejection and pathogen transmission, however in some cases might not be available. For example, in genetic disease suitable autologous cells are not available. Also very ill or elderly persons, as well as patients suffering from severe burns, may not have sufficient quantities of autologous cells to establish useful cell lines. Moreover, since this category of cells needs to be harvested from the patient, there are also some concerns related to the necessity of performing such surgical operations that might lead to donor site infection or chronic pain. Autologous cells also must be cultured from samples before they can be used: this takes time, so autologous solutions may not be very quick. Recently there has been a trend towards the use of mesenchymal stem cells from bone marrow and fat. These cells can differentiate into a variety of tissue types, including bone, cartilage, fat, and nerve. A large number of cells can be easily and quickly isolated from fat, thus opening the potential for large numbers of cells to be quickly and easily obtained.

- Allogeneic cells come from the body of a donor of the same species. While there are some ethical constraints to the use of human cells for in vitro studies, the employment of dermal fibroblasts from human foreskin has been demonstrated to be immunologically safe and thus a viable choice for tissue engineering of skin.

- Xenogenic cells are these isolated from individuals of another species. In particular animal cells have been used quite extensively in experiments aimed at the construction of cardiovascular implants.

- Syngenic or isogenic cells are isolated from genetically identical organisms, such as twins, clones, or highly inbred research animal models.

- Primary cells are from an organism.

- Secondary cells are from a cell bank.

- Stem cells are undifferentiated cells with the ability to divide in culture and give rise to different forms of specialized cells. According to their source stem cells are divided into "adult" and "embryonic" stem cells, the first class being multipotent and the latter mostly pluripotent; some cells are totipotent, in the earliest stages of the embryo. While there is still a large ethical debate related with the use of embryonic stem cells, it is thought that another alternative source - induced stem cells may be useful for the repair of diseased or damaged tissues, or may be used to grow new organs.

Scaffolds

Scaffolds are materials that have been engineered to cause desirable cellular interactions to contribute to the formation of new functional tissues for medical purposes. Cells are often 'seeded' into these structures capable of supporting three-dimensional tissue formation. Scaffolds mimic the extracellular matrix of the native tissue, recapitulating the in vivo milieu and allowing cells to influence their own microenvironments. They usually serve for at least one of the following purposes:

- Allow cell attachment and migration

- Deliver and retain cells and biochemical factors

- Enable diffusion of vital cell nutrients and expressed products

- Exert certain mechanical and biological influences to modify the behaviour of the cell phase

In 2009, an interdisciplinary team led by the thoracic surgeon Thorsten Walles implanted the first bioartificial transplant that provides an innate vascular network for post-transplant graft supply successfully into a patient awaiting tracheal reconstruction.

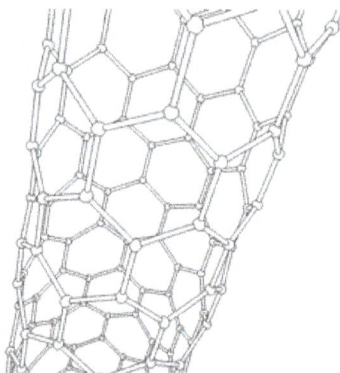

This animation of a rotating carbon nanotube shows its 3D structure. Carbon nanotubes are among the numerous candidates for tissue engineering scaffolds since they are biocompatible, resistant to biodegradation and can be functionalized with biomolecules. However, the possibility of toxicity with non-biodegradable nano-materials is not fully understood.

To achieve the goal of tissue reconstruction, scaffolds must meet some specific requirements. A high porosity and an adequate pore size are necessary to facilitate cell seeding and diffusion throughout the whole structure of both cells and nutrients. Biodegradability is often an essential factor since scaffolds should preferably be absorbed by the surrounding tissues without the necessity of a surgical removal. The rate at which degradation occurs has to coincide as much as possible with the rate of tissue formation: this means that while cells are fabricating their own natural matrix structure around themselves, the scaffold is able to provide structural integrity within the body and eventually it will break down leaving the newly formed tissue which will take over the mechanical load. Injectability is also important for clinical uses. Recent research on organ printing is showing how crucial a good control of the 3D environment is to ensure reproducibility of experiments and offer better results.

Materials

Many different materials (natural and synthetic, biodegradable and permanent) have been investigated. Most of these materials have been known in the medical field before the advent of tissue engineering as a research topic, being already employed as bioresorbable sutures. Examples of these materials are collagen and some polyesters.

New biomaterials have been engineered to have ideal properties and functional customization: injectability, synthetic manufacture, biocompatibility, non-immunogenicity, transparency, nano-scale fibers, low concentration, resorption rates, etc. PuraMatrix, originating from the MIT labs of Zhang, Rich, Grodzinsky and Langer is one of these new biomimetic scaffold families which has now been commercialized and is impacting clinical tissue engineering.

A commonly used synthetic material is PLA - polylactic acid. This is a polyester which degrades within the human body to form lactic acid, a naturally occurring chemical which is easily removed from the body. Similar materials are polyglycolic acid (PGA) and polycaprolactone (PCL): their degradation mechanism is similar to that of PLA, but they exhibit respectively a faster and a slower rate of degradation compared to PLA. While these materials have well maintained mechanical strength and structural integrity, they exhibit a hydrophobic nature. This hydrophobicity inhibits their biocompatibility, which makes them less effective for in vivo use as tissue scaffolding. In order to fix the lack of biocompatibility, much research has been done to combine these hydrophobic materials with hydrophilic and more biocompatible hydrogels. While these hydrogels have a superior biocompatibility, they lack the structural integrity of PLA, PCL, and PGA. By combining the two different types of materials, researchers are trying to create a synergistic relationship that produces a more biocompatible tissue scaffolding. Scaffolds may also be constructed from natural materials: in particular different derivatives of the extracellular matrix have been studied to evaluate their ability to support cell growth. Proteic materials, such as collagen or fibrin, and polysaccharidic materials, like chitosan or glycosaminoglycans (GAGs), have all proved suitable in terms of cell compatibility, but some issues with potential immunogenicity still remains. Among GAGs hyaluronic acid, possibly in combination with cross linking agents (e.g. glutaraldehyde, water-soluble carbodiimide, etc.), is one of the possible choices as scaffold material. Functionalized groups of scaffolds may be useful in the delivery of small molecules (drugs) to specific tissues. Another form of scaffold under investigation is decellularised tissue extracts whereby the remaining cellular remnants/extracellular matrices act as the scaffold. Recently a range of nanocomposites biomaterials are fabricated by incorporating nanomaterials within polymeric matrix to engineer bioactive scaffolds.

A 2009 study by Ratmir et al. aimed to improve in vivo-like conditions for 3D tissue via "stacking and de-stacking layers of paper impregnated with suspensions of cells in extracellular matrix hydrogel, making it possible to control oxygen and nutrient gradients in 3D, and to analyze molecular and genetic responses". It is possible to manipulate gradients of soluble molecules, and to characterize cells in these complex gradients more effectively than conventional 3D cultures based on hydrogels, cell spheroids, or 3D perfusion reactors. Different thicknesses of paper and types of medium can support a variety of experimental environments. Upon deconstruction, these sheets can be useful in cell-based high-throughput screening and drug discovery.

Synthesis

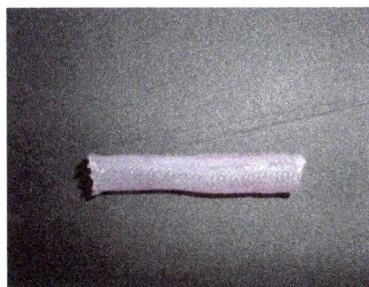

Tissue engineered vascular graft

Tissue engineered heart valve

A number of different methods have been described in literature for preparing porous structures to be employed as tissue engineering scaffolds. Each of these techniques presents its own advantages, but none are free of drawbacks.

Nanofiber self-assembly

Molecular self-assembly is one of the few methods for creating biomaterials with properties similar in scale and chemistry to that of the natural in vivo extracellular matrix (ECM), a crucial step toward tissue engineering of complex tissues. Moreover, these hydrogel scaffolds have shown superiority in in vivo toxicology and biocompatibility compared to traditional macroscaffolds and animal-derived materials.

Textile technologies

These techniques include all the approaches that have been successfully employed for the preparation of non-woven meshes of different polymers. In particular, non-woven polyglycolide structures have been tested for tissue engineering applications: such fibrous structures have been found useful to grow different types of cells. The principal drawbacks are related to the difficulties in obtaining high porosity and regular pore size.

Solvent casting and particulate leaching (SCPL)

This approach allows for the preparation of structures with regular porosity, but with limited thickness. First, the polymer is dissolved into a suitable organic solvent (e.g. polylactic acid could be dissolved into dichloromethane), then the solution is cast into a mold filled with porogen particles. Such porogen can be an inorganic salt like sodium chloride, crystals of saccharose, gelatin spheres or paraffin spheres. The size of the porogen particles will affect the size of the scaffold pores, while the polymer to porogen ratio is directly correlated to the amount of porosity of the final structure. After the polymer solution has been cast the solvent is allowed to fully evaporate, then the composite structure in the mold is immersed in a bath of a liquid suitable for dissolving the porogen: water in the case of sodium chloride, saccharose and gelatin or an aliphatic solvent like hexane for use with paraffin. Once the porogen has been fully dissolved, a porous structure is obtained. Other than the small thickness range that can be obtained, another drawback of SCPL lies in its use of organic solvents which must be fully removed to avoid any possible damage to the cells seeded on the scaffold.

Gas foaming

To overcome the need to use organic solvents and solid porogens, a technique using gas as a porogen has been developed. First, disc-shaped structures made of the desired polymer are prepared by means of compression molding using a heated mold. The discs are then placed in a chamber where they are exposed to high pressure CO_2 for several days. The pressure inside the chamber is gradually restored to atmospheric levels. During this procedure the pores are formed by the carbon dioxide molecules that abandon the polymer, resulting in a sponge-like structure. The main problems resulting from such a technique are caused by the excessive heat used during compression molding (which prohibits the incorporation of any temperature labile material into the polymer matrix) and by the fact that the pores do not form an interconnected structure.

Emulsification/Freeze-drying

This technique does not require the use of a solid porogen like SCPL. First, a synthetic polymer is dissolved into a suitable solvent (e.g. polylactic acid in dichloromethane) then water is added to the polymeric solution and the two liquids are mixed in order to obtain an emulsion. Before the two phases can separate, the emulsion is cast into a mold and quickly frozen by means of immersion into liquid nitrogen. The frozen emulsion is subsequently freeze-dried to remove the dispersed water and the solvent, thus leaving a solidified, porous polymeric structure. While emulsification and freeze-drying allow for a faster preparation when compared to SCPL (since it does not require a time consuming leaching step), it still requires the use of solvents. Moreover, pore size is relatively small and porosity is often irregular. Freeze-drying by itself is also a commonly employed technique for the fabrication of scaffolds. In particular, it is used to prepare collagen sponges: collagen is dissolved into acidic solutions of acetic acid or hydrochloric acid that are cast into a mold, frozen with liquid nitrogen and then lyophilized.

Thermally induced phase separation (TIPS)

Similar to the previous technique, this phase separation procedure requires the use of a solvent with a low melting point that is easy to sublime. For example, dioxane could be used to dissolve polylactic acid, then phase separation is induced through the addition of a small quantity of water: a polymer-rich and a polymer-poor phase are formed. Following cooling below the solvent melting point and some days of vacuum-drying to sublime the solvent, a porous scaffold is obtained. Liquid-liquid phase separation presents the same drawbacks of emulsification/freeze-drying.

Electrospinning

A highly versatile technique that can be used to produce continuous fibers from submicrometer to nanometer diameters. In a typical electrospinning set-up, a solution is fed through a spinneret and a high voltage is applied to the tip. The buildup of electrostatic repulsion within the charged solution, causes it to eject a thin fibrous stream. A mounted collector plate or rod with an opposite or grounded charge draws in the continuous fibers, which arrive to form a highly porous network. The primary advantages of this technique are its simplicity and ease of variation. At a laboratory level, a typical electrospinning set-up only requires a high voltage power supply (up to 30 kV), a syringe, a flat tip needle and a conducting collector. For these reasons, electrospinning has become a common method of scaffold manufacture in many labs. By modifying variables such as the distance to collector, magnitude of applied voltage, or solution flow rate—researchers can dramatically change the overall scaffold architecture.

CAD/CAM technologies

Because most of the above techniques are limited when it comes to the control of porosity and pore size, computer assisted design and manufacturing techniques have been introduced to tissue engineering. First, a three-dimensional structure is designed using CAD software. The porosity can be tailored using algorithms within the software. The scaffold is then realized by using ink-jet printing of polymer powders or through Fused Deposition Modeling of a polymer melt.

A 2011 study by El-Ayoubi et al. investigated "3D-plotting technique to produce (biocompatible and biodegradable) poly-L-Lactide macroporous scaffolds with two different pore sizes" via solid free-form fabrication (SSF) with computer-aided-design (CAD), to explore therapeutic articular cartilage replacement as an "alternative to conventional tissue repair". The study found the smaller the pore size paired with mechanical stress in a bioreactor (to induce in vivo-like conditions), the higher the cell viability in potential therapeutic functionality via decreasing recovery time and increasing transplant effectiveness.

Laser-assisted BioPrinting (LaBP)

In a 2012 study, Koch et al. focused on whether Laser-assisted BioPrinting (LaBP) can be used to build multicellular 3D patterns in natural matrix, and whether the generated constructs are functioning and forming tissue. LaBP arranges small volumes of living cell suspensions in set high-resolution patterns. The investigation was successful, the researchers foresee that "generated tissue constructs might be used for in vivo testing by implanting them into animal models" (14). As of this study, only human skin tissue has been synthesized, though researchers project that "by integrating further cell types (e.g. melanocytes, Schwann cells, hair follicle cells) into the printed cell construct, the behavior of these cells in a 3D in vitro microenvironment similar to their natural one can be analyzed", useful for drug discovery and toxicology studies.

Assembly Methods

One of the continuing, persistent problems with tissue engineering is mass transport limitations. Engineered tissues generally lack an initial blood supply, thus making it difficult for any implanted cells to obtain sufficient oxygen and nutrients to survive, and/or function properly.

Self-assembly

Self-assembly may play an important role here, both from the perspective of encapsulating cells and proteins, as well as creating scaffolds on the right physical scale for engineered tissue constructs and cellular ingrowth. The micromasonry is a prime technology to get cells grown in a lab to assemble into three-dimensional shapes. To break down tissue into single-cell building blocks, researchers have to dissolve the extracellular mortar that normally binds them together. But once that glue is removed, it's quite difficult to get cells to reassemble into the complex structures that make up our natural tissues. While cells aren't easily stackable, building blocks are. So the micromasonry starts with the encapsulation of living cells in polymer cubes. From there, the blocks self-assemble in any shape using templates.

Liquid-based Template Assembly

The air-liquid surface established by Faraday waves is explored as a template to assemble biological entities for bottom-up tissue engineering. This liquid-based template can be dynamically reconfigured in a few seconds, and the assembly on the template can be achieved in a scalable and parallel manner. Assembly of microscale hydrogels, cells, neuron-seeded micro-carrier beads, cell spheroids into various symmetrical and periodic structures was demonstrated with good cell viability. Formation of 3D neural network was achieved after 14-day tissue culture.

Additive Manufacturing

It might be possible to print organs, or possibly entire organisms using additive manufacturing techniques. A recent innovative method of construction uses an ink-jet mechanism to print precise layers of cells in a matrix of thermoreversible gel. Endothelial cells, the cells that line blood vessels, have been printed in a set of stacked rings. When incubated, these fused into a tube.

The field of three-dimensional and highly accurate models of biological systems is pioneered by multiple projects and technologies including a rapid method for creating tissues and even whole organs involves a 3D printer that can print the scaffolding and cells layer by layer into a working tissue sample or organ. The device is presented in a TED talk by Dr. Anthony Atala, M.D. the Director of the Wake Forest Institute for Regenerative Medicine, and the W.H. Boyce Professor and Chair of the Department of Urology at Wake Forest University, in which a kidney is printed on stage during the seminar and then presented to the crowd. It is anticipated that this technology will enable the production of livers in the future for transplantation and theoretically for toxicology and other biological studies as well.

Recently Multi-Photon Processing (MPP) was employed for in vivo expperiments by engineering artificial cartilage constructs. An ex vivo histological examination showed that certain pore geometry and the pre-growing of chondrocytes (Cho) prior to implantation significantly improves the performance of the created 3D scaffolds. The achieved biocompatibility was comparable to the commercially available collagen membranes. The successful outcome of this study supports the idea that hexagonal-pore-shaped hybrid organic-inorganic microstructured scaffolds in combination with Cho seeding may be successfully implemented for cartilage tissue engineering.

Scaffolding

In 2013, using a 3-d scaffolding of Matrigel in various configurations, substantial pancreatic organoids was produced in vitro. Clusters of small numbers of cells proliferated into 40,000 cells within one week. The clusters transform into cells that make either digestive enzymes or hormones like insulin, self-organizing into branched pancreatic organoids that resemble the pancreas.

The cells are sensitive to the environment, such as gel stiffness and contact with other cells. Individual cells do not thrive; a minimum of four proximate cells was required for subsequent organoid development. Modifications to the medium composition produced either hollow spheres mainly composed of pancreatic progenitors, or complex organoids that spontaneously undergo pancreatic morphogenesis and differentiation. Maintenance and expansion of pancreatic progenitors require active Notch and FGF signaling, recapitulating in vivo niche signaling interactions.

The organoids were seen as potentially offering mini-organs for drug testing and for spare insulin-producing cells.

Tissue Culture

In many cases, creation of functional tissues and biological structures in vitro requires extensive culturing to promote survival, growth and inducement of functionality. In general, the basic requirements of cells must be maintained in culture, which include oxygen, pH, humidity, temperature, nutrients and osmotic pressure maintenance.

Tissue engineered cultures also present additional problems in maintaining culture conditions. In standard cell culture, diffusion is often the sole means of nutrient and metabolite transport. However, as a culture becomes larger and more complex, such as the case with engineered organs and whole tissues, other mechanisms must be employed to maintain the culture, such as the creation of capillary networks within the tissue.

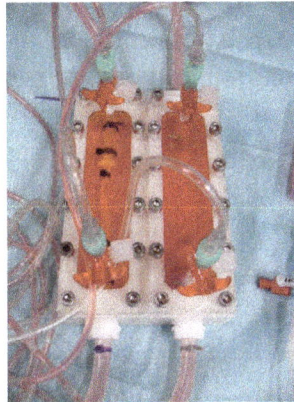
Bioreactor for cultivation of vascular grafts

Another issue with tissue culture is introducing the proper factors or stimuli required to induce functionality. In many cases, simple maintenance culture is not sufficient. Growth factors, hormones, specific metabolites or nutrients, chemical and physical stimuli are sometimes required. For example, certain cells respond to changes in oxygen tension as part of their normal development, such as chondrocytes, which must adapt to low oxygen conditions or hypoxia during skeletal development. Others, such as endothelial cells, respond to shear stress from fluid flow, which is encountered in blood vessels. Mechanical stimuli, such as pressure pulses seem to be beneficial to all kind of cardiovascular tissue such as heart valves, blood vessels or pericardium.

Bioreactors

A bioreactor in tissue engineering, as opposed to industrial bioreactors, is a device that attempts to simulate a physiological environment in order to promote cell or tissue growth in vitro. A physiological environment can consist of many different parameters such as temperature and oxygen or carbon dioxide concentration, but can extend to all kinds of biological, chemical or mechanical stimuli. Therefore, there are systems that may include the application of forces or stresses to the tissue or even of electric current in two- or three-dimensional setups.

In academic and industry research facilities, it is typical for bioreactors to be developed to replicate the specific physiological environment of the tissue being grown (e.g., flex and fluid shearing for heart tissue growth). Several general-use and application-specific bioreactors are also commercially available, and may provide static chemical stimulation or combination of chemical and mechanical stimulation.

The Bioreactors used for 3D cell cultures are small plastic cylindrical chambers with regulated internal humidity and moisture specifically engineered for the purpose of growing cells in three dimensions. The bioreactor uses bioactive synthetic materials such as polyethylene terephthalate membranes to surround the spheroid cells in an environment that maintains high levels of nutrients. They are easy to open and close, so that cell spheroids can be removed for testing, yet the chamber is able to maintain 100% humidity throughout. This humidity is important to achieve maximum cell growth and function. The bioreactor chamber is part of a larger device that rotates to ensure equal cell growth in each direction across three dimensions. MC2 Biotek has developed a bioreactor known as ProtoTissue that uses gas exchange to maintain high oxygen levels within the cell chamber; improving upon previous bioreactors, because the higher oxygen levels help the cell grow and undergo normal cell respiration.

Long Fiber Generation

In 2013, a group from the University of Tokyo developed cell laden fibers up to a meter in length and on the order of 100 μm in size. These fibers were created using a microfluidic device that forms a double coaxial laminar flow. Each 'layer' of the microfluidic device (cells seeded in ECM, a hydrogel sheath, and finally a calcium chloride solution). The seeded cells culture within the hydrogel sheath for several days, and then the sheath is removed with viable cell fibers. Various cell types were inserted into the ECM core, including myocytes, endothelial cells, nerve cell fibers, and epithelial cell fibers. This group then showed that these fibers can be woven together to fabricate tissues or organs in a mechanism similar to textile weaving. Fibrous morphologies are advantageous in that they provide an alternative to traditional scaffold design, and many organs (such as muscle) are composed of fibrous cells.

Bioartificial Organs

An artificial organ is a man-made device that is implanted or integrated into a human to replace a natural organ, for the purpose of restoring a specific function or a group of related functions so the patient may return to a normal life as soon as possible. The replaced function doesn't necessarily have to be related to life support, but often is. The ultimate goal of tissue engineering as a discipline is to allow both 'off the shelf' bioartificial organs and regeneration of injured tissue in the body. In order to successfully create bioartificial organs from a patients stem cells, researchers continue to make improvements in the generation of complex tissues by tissue engineering. For example, much research is aimed at understanding nanoscale cues present in a cell's microenvironment.

Ceramic Engineering

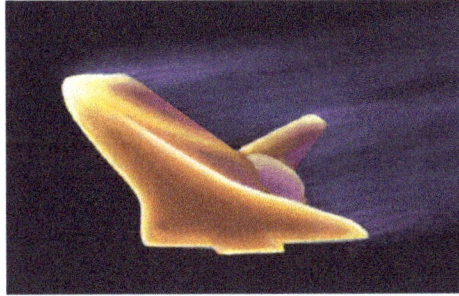

Simulation of the outside of the Space Shuttle as it heats up to over 1,500 °C (2,730 °F) during re-entry into the Earth's atmosphere

Bearing components made from 100% silicon nitride Si_3N_4

Ceramic bread knife

Ceramic engineering is the science and technology of creating objects from inorganic, non-metallic materials. This is done either by the action of heat, or at lower temperatures using precipitation reactions from high-purity chemical solutions. The term includes the purification of raw materials, the study and production of the chemical compounds concerned, their formation into components and the study of their structure, composition and properties.

Ceramic materials may have a crystalline or partly crystalline structure, with long-range order on atomic scale. Glass ceramics may have an amorphous or glassy structure, with limited or short-range atomic order. They are either formed from a molten mass that solidifies on cooling, formed and matured by the action of heat, or chemically synthesized at low temperatures using, for example, hydrothermal or sol-gel synthesis.

The special character of ceramic materials gives rise to many applications in materials engineering, electrical engineering, chemical engineering and mechanical engineering. As ceramics are heat resistant, they can be used for many tasks for which materials like metal and polymers are unsuitable. Ceramic materials are used in a wide range of industries, including mining, aerospace, medicine, refinery, food and chemical industries, packaging science, electronics, industrial and transmission electricity, and guided lightwave transmission.

History

It is related to the older Indo-European language root "to burn", "Ceramic" may be used as a noun in the singular to refer to a ceramic material or the product of ceramic manufacture, or as an adjective. The plural "ceramics" may be used to refer the making of things out of ceramic materials. Ceramic engineering, like many sciences, evolved from a different discipline by today's standards. Materials science engineering is grouped with ceramics engineering to this day.

Leo Morandi's tile glazing line (circa 1945)

Abraham Darby first used coke in 1709 in Shropshire, England, to improve the yield of a smelting process. Coke is now widely used to produce carbide ceramics. Potter Josiah Wedgwood opened the first modern ceramics factory in Stoke-on-Trent, England, in 1759. Austrian chemist Carl Josef Bayer, working for the textile industry in Russia, developed a process to separate alumina from bauxite ore in 1888. The Bayer process is still used to purify alumina for the ceramic and aluminium industries. Brothers Pierre and Jacques Curie discovered piezoelectricity in Rochelle salt circa 1880. Piezoelectricity is one of the key properties of electroceramics.

E.G. Acheson heated a mixture of coke and clay in 1893, and invented carborundum, or synthetic silicon carbide. Henri Moissan also synthesized SiC and tungsten carbide in his electric arc furnace in Paris about the same time as Acheson. Karl Schröter used liquid-phase sintering to bond or "cement" Moissan's tungsten carbide particles with cobalt in 1923 in Germany. Cemented (metal-bonded) carbide edges greatly increase the durability of hardened steel cutting tools. W.H. Nernst developed cubic-stabilized zirconia in the 1920s in Berlin. This material is used as an oxygen sensor in exhaust systems. The main limitation on the use of ceramics in engineering is brittleness.

Military

Soldiers pictured during the 2003 Iraq War seen through IR transparent Night Vision Goggles

The military requirements of World War II encouraged developments, which created a need for high-performance materials and helped speed the development of ceramic science and engineering. Throughout the 1960s and 1970s, new types of ceramics were developed in response to advances in atomic energy, electronics, communications, and space travel. The discovery of ceramic superconductors in 1986 has spurred intense research to develop superconducting ceramic parts for electronic devices, electric motors, and transportation equipment.

There is an increasing need in the military sector for high-strength, robust materials which have the capability to transmit light around the visible (0.4–0.7 micrometers) and mid-infrared (1–5 micrometers) regions of the spectrum. These materials are needed for applications requiring transparent armour. Transparent armor is a material or system of materials designed to be optically transparent, yet protect from fragmentation or ballistic impacts. The primary requirement for a transparent armour system is to not only defeat the designated threat but also provide a multi-hit capability with minimized distortion of surrounding areas. Transparent armour windows must also be compatible with night vision equipment. New materials that are thinner, lightweight, and offer better ballistic performance are being sought. Such solid-state components have found widespread use for various applications in the electro-optical field including: optical fibres for guided lightwave transmission, optical switches, laser amplifiers and lenses, hosts for solid-state lasers and optical window materials for gas lasers, and infrared (IR) heat seeking devices for missile guidance systems and IR night vision.

Modern Industry

Now a multibillion-dollar a year industry, ceramic engineering and research has established itself as an important field of science. Applications continue to expand as researchers develop new kinds of ceramics to serve different purposes.

- Zirconium dioxide ceramics are used in the manufacture of knives. The blade of the ceramic knife will stay sharp for much longer than that of a steel knife, although it is more brittle and can be snapped by dropping it on a hard surface.

- Ceramics such as alumina, boron carbide and silicon carbide have been used in bulletproof vests to repel small arms rifle fire. Such plates are known commonly as trauma plates. Similar material is used to protect cockpits of some military aircraft, because of the low weight of the material.

- Silicon nitride parts are used in ceramic ball bearings. Their higher hardness means that they are much less susceptible to wear and can offer more than triple lifetimes. They also deform less under load meaning they have less contact with the bearing retainer walls and can roll faster. In very high speed applications, heat from friction during rolling can cause problems for metal bearings; problems which are reduced by the use of ceramics. Ceramics are also more chemically resistant and can be used in wet environments where steel bearings would rust. The major drawback to using ceramics is a significantly higher cost. In many cases their electrically insulating properties may also be valuable in bearings.

- In the early 1980s, Toyota researched production of an adiabatic ceramic engine which can run at a temperature of over 6000 °F (3300 °C). Ceramic engines do not require a cooling system and hence allow a major weight reduction and therefore greater fuel efficiency. Fuel efficiency of the engine is also higher at high temperature, as shown by Carnot's theorem. In a conventional metallic engine, much of the energy released from the fuel must be dissipated as waste heat in order to prevent a meltdown of the metallic parts. Despite all of these desirable properties, such engines are not in production because the manufacturing of ceramic parts in the requisite precision and durability is difficult. Imperfection in the ceramic leads to cracks, which can lead to potentially dangerous equipment failure. Such engines are possible in laboratory settings, but mass-production is not feasible with current technology.

- Work is being done in developing ceramic parts for gas turbine engines. Currently, even blades made of advanced metal alloys used in the engines' hot section require cooling and careful limiting of operating temperatures. Turbine engines made with ceramics could operate more efficiently, giving aircraft greater range and payload for a set amount of fuel.

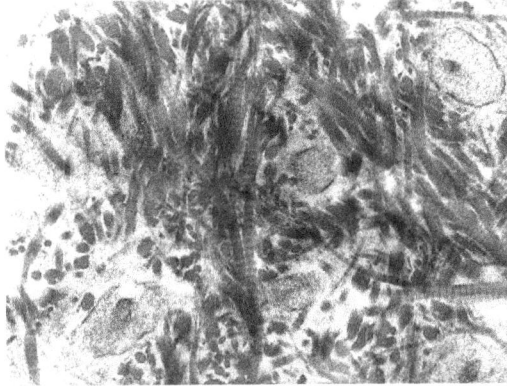

Collagen fibers of woven bone

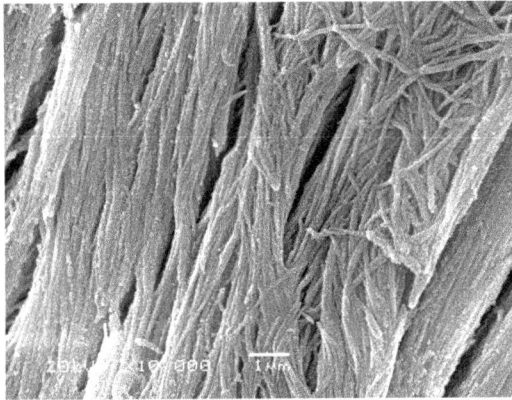

Scanning electron microscopy image of bone

- Recently, there have been advances in ceramics which include bio-ceramics, such as dental implants and synthetic bones. Hydroxyapatite, the natural mineral component of bone, has been made synthetically from a number of biological and chemical sources and can be formed into ceramic materials. Orthopaedic implants made from these materials bond readily to bone and other tissues in the body without rejection or inflammatory reactions. Because of this, they are of great interest for gene delivery and tissue engineering scaffolds. Most hydroxyapatite ceramics are very porous and lack mechanical strength and are used to coat metal orthopaedic devices to aid in forming a bond to bone or as bone fillers. They are also used as fillers for orthopaedic plastic screws to aid in reducing the inflammation and increase absorption of these plastic materials. Work is being done to make strong, fully dense nano crystalline hydroxyapatite ceramic materials for orthopaedic weight bearing devices, replacing foreign metal and plastic orthopaedic materials with a synthetic, but naturally occurring, bone mineral. Ultimately these ceramic materials may be used as bone replacements or with the incorporation of protein collagens, synthetic bones.

- High-tech ceramic is used in watch-making for producing watch cases. The material is valued by watchmakers for its light weight, scratch-resistance, durability and smooth touch. IWC is one of the brands that initiated the use of ceramic in watch-making. The case of the IWC 2007 Top Gun edition of the Pilot's Watch Double chronograph is crafted in high-tech black ceramic.

Glass-ceramics

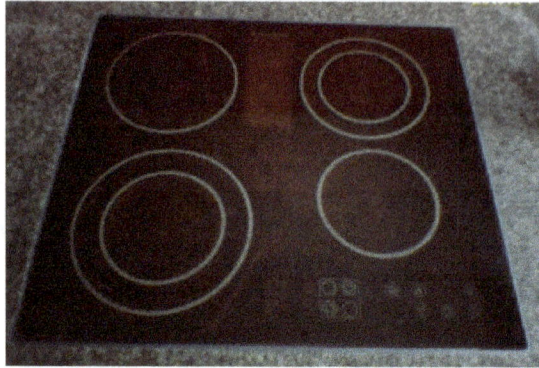

A high strength glass-ceramic cook-top with negligible thermal expansion.

Glass-ceramic materials share many properties with both glasses and ceramics. Glass-ceramics have an amorphous phase and one or more crystalline phases and are produced by a so-called "controlled crystallization", which is typically avoided in glass manufacturing. Glass-ceramics often contain a crystalline phase which constitutes anywhere from 30% [m/m] to 90% [m/m] of its composition by volume, yielding an array of materials with interesting thermomechanical properties.

In the processing of glass-ceramics, molten glass is cooled down gradually before re-heating and annealing. In this heat treatment the glass partly crystallizes. In many cases, so-called 'nucleation agents' are added in order to regulate and control the crystallization process. Because there is usually no pressing and sintering, glass-ceramics do not contain the volume fraction of porosity typically present in sintered ceramics.

The term mainly refers to a mix of lithium and aluminosilicates which yields an array of materials with interesting thermomechanical properties. The most commercially important of these have the distinction of being impervious to thermal shock. Thus, glass-ceramics have become extremely useful for countertop cooking. The negative thermal expansion coefficient (TEC) of the crystalline ceramic phase can be balanced with the positive TEC of the glassy phase. At a certain point (~70% crystalline) the glass-ceramic has a net TEC near zero. This type of glass-ceramic exhibits excellent mechanical properties and can sustain repeated and quick temperature changes up to 1000 °C.

Processing Steps

The traditional ceramic process generally follows this sequence: Milling → Batching → Mixing → Forming → Drying → Firing → Assembly.

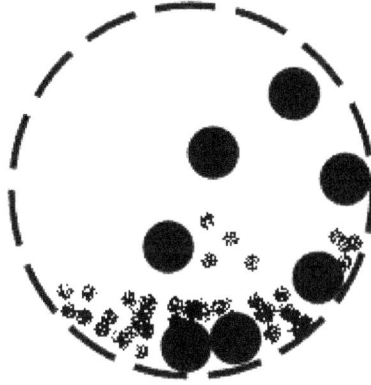

Ball mill

- Milling is the process by which materials are reduced from a large size to a smaller size. Milling may involve breaking up cemented material (in which case individual particles retain their shape) or pulverization (which involves grinding the particles themselves to a smaller size). Milling is generally done by mechanical means, including attrition (which is particle-to-particle collision that results in agglomerate break up or particle shearing), compression (which applies a forces that results in fracturing), and impact (which employs a milling medium or the particles themselves to cause fracturing). Attrition milling equipment includes the wet scrubber (also called the planetary mill or wet attrition mill), which has paddles in water creating vortexes in which the material collides and break up. Compression mills include the jaw crusher, roller crusher and cone crusher. Impact mills include the ball mill, which has media that tumble and fracture the material. Shaft impactors cause particle-to particle attrition and compression.

- Batching is the process of weighing the oxides according to recipes, and preparing them for mixing and drying.

- Mixing occurs after batching and is performed with various machines, such as dry mixing ribbon mixers (a type of cement mixer), Mueller mixers,[clarification needed] and pug mills. Wet mixing generally involves the same equipment.

- Forming is making the mixed material into shapes, ranging from toilet bowls to spark plug insulators. Forming can involve: (1) Extrusion, such as extruding "slugs" to make bricks, (2) Pressing to make shaped parts, (3) Slip casting, as in making toilet bowls, wash basins and ornamentals like ceramic statues. Forming produces a "green" part, ready for drying. Green parts are soft, pliable, and over time will lose shape. Handling the green product will change its shape. For example, a green brick can be "squeezed", and after squeezing it will stay that way.

- Drying is removing the water or binder from the formed material. Spray drying is widely used to prepare powder for pressing operations. Other dryers are tunnel dryers and periodic dryers. Controlled heat is applied in this two-stage process. First, heat removes water. This step needs careful control, as rapid heating causes cracks and surface defects. The dried part is smaller than the green part, and is brittle, necessitating careful handling, since a small impact will cause crumbling and breaking.

- Sintering is where the dried parts pass through a controlled heating process, and the oxides are chemically changed to cause bonding and densification. The fired part will be smaller than the dried part.

Forming Methods

Ceramic forming techniques include throwing, slipcasting, tape casting, freeze-casting, injection moulding, dry pressing, isostatic pressing, hot isostatic pressing (HIP) and others. Methods for forming ceramic powders into complex shapes are desirable in many areas of technology. Such methods are required for producing advanced, high-temperature structural parts such as heat engine components and turbines. Materials other than ceramics which are used in these processes may include: wood, metal, water, plaster and epoxy—most of which will be eliminated upon firing.

These forming techniques are well known for providing tools and other components with dimensional stability, surface quality, high (near theoretical) density and microstructural uniformity. The increasing use and diversity of speciality forms of ceramics adds to the diversity of process technologies to be used.

Thus, reinforcing fibres and filaments are mainly made by polymer, sol-gel, or CVD processes, but melt processing also has applicability. The most widely used speciality form is layered structures, with tape casting for electronic substrates and packages being pre-eminent. Photo-lithography is of increasing interest for precise patterning of conductors and other components for such packaging. Tape casting or forming processes are also of increasing interest for other applications, ranging from open structures such as fuel cells to ceramic composites.

The other major layer structure is coating, where melt spraying is very important, but chemical and physical vapour deposition and chemical (e.g., sol-gel and polymer pyrolysis) methods are all seeing increased use. Besides open structures from formed tape, extruded structures, such as honeycomb catalyst supports, and highly porous structures, including various foams, for example, reticulated foam, are of increasing use.

Densification of consolidated powder bodies continues to be achieved predominantly by (pressureless) sintering. However, the use of pressure sintering by hot pressing is increasing, especially for non-oxides and parts of simple shapes where higher quality (mainly microstructural homogeneity) is needed, and larger size or multiple parts per pressing can be an advantage.

The Sintering Process

The principles of sintering-based methods are simple ("sinter" has roots in the English "cinder"). The firing is done at a temperature below the melting point of the ceramic. Once a roughly-held-together object called a "green body" is made, it is baked in a kiln, where atomic and molecular diffusion processes give rise to significant changes in the primary microstructural features. This includes the gradual elimination of porosity, which is typically accompanied by a net shrinkage and overall densification of the component. Thus, the pores in the object may close up, resulting in a denser product of significantly greater strength and fracture toughness.

Another major change in the body during the firing or sintering process will be the establishment of the polycrystalline nature of the solid. This change will introduce some form of grain size distribution, which will have a significant impact on the ultimate physical properties of the material. The grain sizes will either be associated with the initial particle size, or possibly the sizes of aggregates or particle clusters which arise during the initial stages of processing.

The ultimate microstructure (and thus the physical properties) of the final product will be limited by and subject to the form of the structural template or precursor which is created in the initial stages of chemical synthesis and physical forming. Hence the importance of chemical powder and polymer processing as it pertains to the synthesis of industrial ceramics, glasses and glass-ceramics.

There are numerous possible refinements of the sintering process. Some of the most common involve pressing the green body to give the densification a head start and reduce the sintering time needed. Sometimes organic binders such as polyvinyl alcohol are added to hold the green body together; these burn out during the firing (at 200–350 °C). Sometimes organic lubricants are added during pressing to increase densification. It is common to combine these, and add binders and lubricants to a powder, then press. (The formulation of these organic chemical additives is an art in itself. This is particularly important in the manufacture of high performance ceramics such as those used by the billions for electronics, in capacitors, inductors, sensors, etc.)

A slurry can be used in place of a powder, and then cast into a desired shape, dried and then sintered. Indeed, traditional pottery is done with this type of method, using a plastic mixture worked with the hands. If a mixture of different materials is used together in a ceramic, the sintering temperature is sometimes above the melting point of one minor component – a liquid phase sintering. This results in shorter sintering times compared to solid state sintering.

Strength of Ceramics

A material's strength is dependent on its microstructure. The engineering processes to which a material is subjected can alter its microstructure. The variety of strengthening mechanisms that alter the strength of a material include the mechanism of grain boundary strengthening. Thus, although yield strength is maximized with decreasing grain size, ultimately, very small grain sizes make the material brittle. Considered in tandem with the fact that the yield strength is the parameter that predicts plastic deformation in the material, one can make informed decisions on how to increase the strength of a material depending on its microstructural properties and the desired end effect.

The relation between yield stress and grain size is described mathematically by the Hall-Petch equation which is

$$\sigma_y = \sigma_0 + \frac{k_y}{\sqrt{d}}$$

where k_y is the strengthening coefficient (a constant unique to each material), σ_o is a materials constant for the starting stress for dislocation movement (or the resistance of the lattice to dislocation motion), d is the grain diameter, and σ_y is the yield stress.

Theoretically, a material could be made infinitely strong if the grains are made infinitely small. This is, unfortunately, impossible because the lower limit of grain size is a single unit cell of the material. Even then, if the grains of a material are the size of a single unit cell, then the material is in fact amorphous, not crystalline, since there is no long range order, and dislocations can not be defined in an amorphous material. It has been observed experimentally that the microstructure with the highest yield strength is a grain size of about 10 nanometres, because grains smaller than this undergo another yielding mechanism, grain boundary sliding. Producing engineering materials with this ideal grain size is difficult because of the limitations of initial particle sizes inherent to nanomaterials and nanotechnology.

Theory of Chemical Processing

Microstructural Uniformity

In the processing of fine ceramics, the irregular particle sizes and shapes in a typical powder often lead to non-uniform packing morphologies that result in packing density variations in the powder compact. Uncontrolled agglomeration of powders due to attractive van der Waals forces can also give rise to in microstructural inhomogeneities.

Differential stresses that develop as a result of non-uniform drying shrinkage are directly related to the rate at which the solvent can be removed, and thus highly dependent upon the distribution of porosity. Such stresses have been associated with a plastic-to-brittle transition in consolidated bodies, and can yield to crack propagation in the unfired body if not relieved.

In addition, any fluctuations in packing density in the compact as it is prepared for the kiln are often amplified during the sintering process, yielding inhomogeneous densification. Some pores and other structural defects associated with density variations have been shown to play a detrimental role in the sintering process by growing and thus limiting end-point densities. Differential stresses arising from inhomogeneous densification have also been shown to result in the propagation of internal cracks, thus becoming the strength-controlling flaws.

It would therefore appear desirable to process a material in such a way that it is physically uniform with regard to the distribution of components and porosity, rather than using particle size distributions which will maximize the green density. The containment of a uniformly dispersed assembly of strongly interacting particles in suspension requires total control over particle-particle interactions. Monodisperse colloids provide this potential.

Monodisperse powders of colloidal silica, for example, may therefore be stabilized sufficiently to ensure a high degree of order in the colloidal crystal or polycrystalline colloidal solid which results from aggregation. The degree of order appears to be limited by the time and space allowed for longer-range correlations to be established.

Such defective polycrystalline colloidal structures would appear to be the basic elements of submicrometer colloidal materials science, and, therefore, provide the first step in developing a more rigorous understanding of the mechanisms involved in microstructural evolution in inorganic systems such as polycrystalline ceramics.

Self-assembly

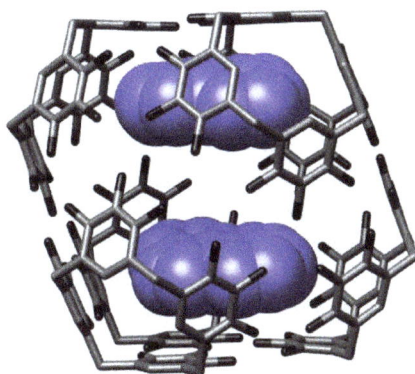

An example of a supramolecular assembly.

Self-assembly is the most common term in use in the modern scientific community to describe the spontaneous aggregation of particles (atoms, molecules, colloids, micelles, etc.) without the influence of any external forces. Large groups of such particles are known to assemble themselves into thermodynamically stable, structurally well-defined arrays, quite reminiscent of one of the 7 crystal systems found in metallurgy and mineralogy (e.g. face-centred cubic, body-centred cubic, etc.). The fundamental difference in equilibrium structure is in the spatial scale of the unit cell (or lattice parameter) in each particular case.

Thus, self-assembly is emerging as a new strategy in chemical synthesis and nanotechnology. Molecular self-assembly has been observed in various biological systems and underlies the formation of a wide variety of complex biological structures. Molecular crystals, liquid crystals, colloids, micelles, emulsions, phase-separated polymers, thin films and self-assembled monolayers all represent examples of the types of highly ordered structures which are obtained using these techniques. The distinguishing feature of these methods is self-organization in the absence of any external forces.

In addition, the principal mechanical characteristics and structures of biological ceramics, polymer composites, elastomers, and cellular materials are being re-evaluated, with an emphasis on bioinspired materials and structures. Traditional approaches focus on design methods of biological materials using conventional synthetic materials. This includes an emerging class of mechanically superior biomaterials based on microstructural features and designs found in nature. The new horizons have been identified in the synthesis of bioinspired materials through processes that are characteristic of biological systems in nature. This includes the nanoscale self-assembly of the components and the development of hierarchical structures.

Ceramic Composites

The Porsche Carrera GT's carbon-ceramic (silicon carbide) composite disc brake

Substantial interest has arisen in recent years in fabricating ceramic composites. While there is considerable interest in composites with one or more non-ceramic constituents, the greatest attention is on composites in which all constituents are ceramic. These typically comprise two ceramic constituents: a continuous matrix, and a dispersed phase of ceramic particles, whiskers, or short (chopped) or continuous ceramic fibres. The challenge, as in wet chemical processing, is to obtain a uniform or homogeneous distribution of the dispersed particle or fibre phase.

Consider first the processing of particulate composites. The particulate phase of greatest interest is tetragonal zirconia because of the toughening that can be achieved from the phase transformation from the metastable tetragonal to the monoclinic crystalline phase, aka transformation toughening. There is also substantial interest in dispersion of hard, non-oxide phases such as SiC, TiB, TiC, boron, carbon and especially oxide matrices like alumina and mullite. There is also interest too incorporating other ceramic particulates, especially those of highly anisotropic thermal expansion. Examples include Al_2O_3, TiO_2, graphite, and boron nitride.

Silicon carbide single crystal

In processing particulate composites, the issue is not only homogeneity of the size and spatial distribution of the dispersed and matrix phases, but also control of the matrix grain size. However, there is some built-in self-control due to inhibition of matrix grain growth by the dispersed phase. Particulate composites, though generally offer increased resistance to damage, failure, or both, are still quite sensitive to inhomogeneities of composition as well as other processing defects such as pores. Thus they need good processing to be effective.

Particulate composites have been made on a commercial basis by simply mixing powders of the two constituents. Although this approach is inherently limited in the homogeneity that can be achieved, it is the most readily adaptable for existing ceramic production technology. However, other approaches are of interest.

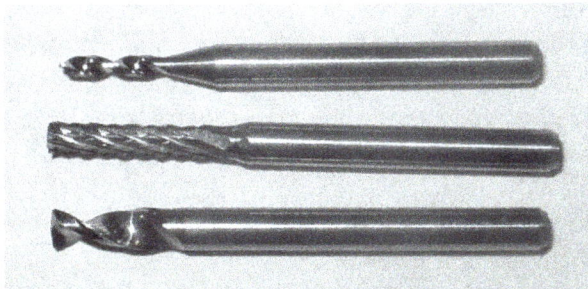

Tungsten carbide milling bits

From the technological standpoint, a particularly desirable approach to fabricating particulate composites is to coat the matrix or its precursor onto fine particles of the dispersed phase with good control of the starting dispersed particle size and the resultant matrix coating thickness. One should in principle be able to achieve the ultimate in homogeneity of distribution and thereby optimize composite performance. This can also have other ramifications, such as allowing more useful composite performance to be achieved in a body having porosity, which might be desired for other factors, such as limiting thermal conductivity.

There are also some opportunities to utilize melt processing for fabrication of ceramic, particulate, whisker and short-fibre, and continuous-fibre composites. Clearly, both particulate and whisker composites are conceivable by solid-state precipitation after solidification of the melt. This can also be obtained in some cases by sintering, as for precipitation-toughened, partially stabilized zirconia. Similarly, it is known that one can directionally solidify ceramic eutectic mixtures and hence obtain uniaxially aligned fibre composites. Such composite processing has typically been limited to very simple shapes and thus suffers from serious economic problems due to high machining costs.

Clearly, there are possibilities of using melt casting for many of these approaches. Potentially even more desirable is using melt-derived particles. In this method, quenching is done in a solid solution or in a fine eutectic structure, in which the particles are then processed by more typical ceramic powder processing methods into a useful body. There have also been preliminary attempts to use melt spraying as a means of forming composites by introducing the dispersed particulate, whisker, or fibre phase in conjunction with the melt spraying process.

Other methods besides melt infiltration to manufacture ceramic composites with long fibre reinforcement are chemical vapour infiltration and the infiltration of fibre preforms with organic precursor, which after pyrolysis yield an amorphous ceramic matrix, initially with a low density. With repeated cycles of infiltration and pyrolysis one of those types of ceramic matrix composites is produced. Chemical vapour infiltration is used to manufacture carbon/carbon and silicon carbide reinforced with carbon or silicon carbide fibres.

Besides many process improvements, the first of two major needs for fibre composites is lower fibre costs. The second major need is fibre compositions or coatings, or composite processing, to reduce degradation that results from high-temperature composite exposure under oxidizing conditions.

Applications

The products of technical ceramics include tiles used in the Space Shuttle program, gas burner nozzles, ballistic protection, nuclear fuel uranium oxide pellets, bio-medical implants, jet engine turbine blades, and missile nose cones.

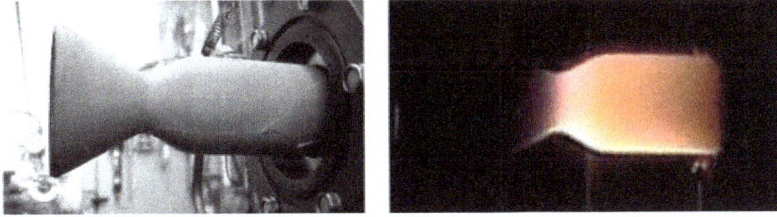

Silicon nitride thruster. Left: Mounted in test stand. Right: Being tested with H_2/O_2 propellants

Its products are often made from materials other than clay, chosen for their particular physical properties. These may be classified as follows:

- Oxides: silica, alumina, zirconia

- Non-oxides: carbides, borides, nitrides, silicides

- Composites: particulate or whisker reinforced matrices, combinations of oxides and non-oxides (e.g. polymers).

Ceramics can be used in many technological industries. One application is the ceramic tiles on NASA's Space Shuttle, used to protect it and the future supersonic space planes from the searing heat of re-entry into the Earth's atmosphere. They are also used widely in electronics and optics. In addition to the applications listed here, ceramics are also used as a coating in various engineering cases. An example would be a ceramic bearing coating over a titanium frame used for an aircraft. Recently the field has come to include the studies of single crystals or glass fibres, in addition to traditional polycrystalline materials, and the applications of these have been overlapping and changing rapidly.

Aerospace

- Engines; Shielding a hot running aircraft engine from damaging other components.

- Airframes; Used as a high-stress, high-temp and lightweight bearing and structural component.

- Missile nose-cones; Shielding the missile internals from heat.

- Space Shuttle tiles

- Space-debris ballistic shields – ceramic fiber woven shields offer better protection to hypervelocity (~7 km/s) particles than aluminium shields of equal weight.

- Rocket nozzles, withstands and focuses the exhaust of the rocket booster.

- Unmanned Air Vehicles; Implications of ceramic engine utilization in aeronautical applications (such as Unmanned Air Vehicles) may result in enhanced performance characteristics and less operational costs.

Biomedical

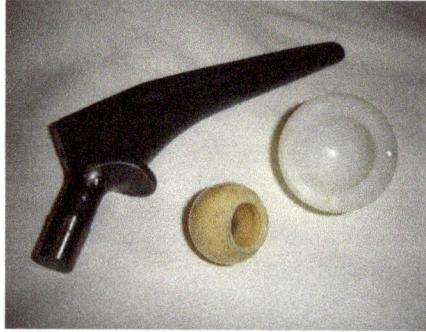

A titanium hip prosthesis, with a ceramic head and polyethylene acetabular cup.

- Artificial bone; Dentistry applications, teeth.
- Biodegradable splints; Reinforcing bones recovering from osteoporosis
- Implant material

Electronics

- Capacitors
- Integrated circuit packages
- Transducers
- Insulators

Optical

- Optical fibres, guided lightwave transmission
- Switches
- Laser amplifiers
- Lenses
- Infrared heat-seeking devices

Automotive

- Heat shield
- Exhaust heat management

Biomaterials

Silicification is quite common in the biological world and occurs in bacteria, single-celled organisms, plants, and animals (invertebrates and vertebrates). Crystalline minerals formed in such environment often show exceptional physical properties (e.g. strength, hardness, fracture toughness) and tend to form hierarchical structures that exhibit microstructural order over a range of length or spatial scales. The minerals are crystallized from an environment that is undersaturated with respect to silicon, and under conditions of neutral pH and low temperature (0–40 °C). Formation of the mineral may occur either within or outside of the cell wall of an organism, and specific biochemical reactions for mineral deposition exist that include lipids, proteins and carbohydrates.

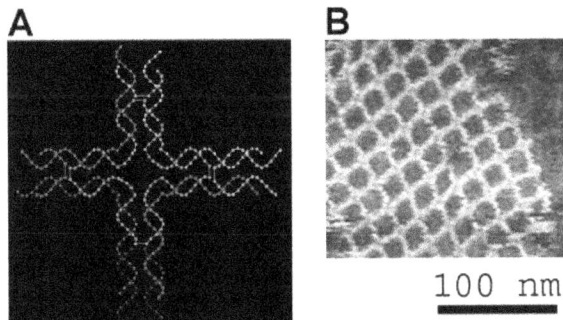

The DNA structure at left (schematic shown) will self-assemble into the structure visualized by atomic force microscopy at right.

Most natural (or biological) materials are complex composites whose mechanical properties are often outstanding, considering the weak constituents from which they are assembled. These complex structures, which have risen from hundreds of million years of evolution, are inspiring the design of novel materials with exceptional physical properties for high performance in adverse conditions. Their defining characteristics such as hierarchy, multifunctionality, and the capacity for self-healing, are currently being investigated.

The basic building blocks begin with the 20 amino acids and proceed to polypeptides, polysaccharides, and polypeptides–saccharides. These, in turn, compose the basic proteins, which are the primary constituents of the 'soft tissues' common to most biominerals. With well over 1000 proteins possible, current research emphasizes the use of collagen, chitin, keratin, and elastin. The 'hard' phases are often strengthened by crystalline minerals, which nucleate and grow in a biomediated environment that determines the size, shape and distribution of individual crystals. The most important mineral phases have been identified as hydroxyapatite, silica, and aragonite. Using the classification of Wegst and Ashby, the principal mechanical characteristics and structures of biological ceramics, polymer composites, elastomers, and cellular materials have been presented. Selected systems in each class are being investigated with emphasis on the relationship between their microstructure over a range of length scales and their mechanical response.

Thus, the crystallization of inorganic materials in nature generally occurs at ambient temperature and pressure. Yet the vital organisms through which these minerals form are capable of consistently producing extremely precise and complex structures. Understanding the processes in which living organisms control the growth of crystalline minerals such as silica could lead to significant advances in the field of materials science, and open the door to novel synthesis techniques for nanoscale composite materials, or nanocomposites.

The iridescent nacre inside a Nautilus shell.

High-resolution SEM observations were performed of the microstructure of the mother-of-pearl (or nacre) portion of the abalone shell. Those shells exhibit the highest mechanical strength and fracture toughness of any non-metallic substance known. The nacre from the shell of the abalone has become one of the more intensively studied biological structures in materials science. Clearly visible in these images are the neatly stacked (or ordered) mineral tiles separated by thin organic sheets along with a macrostructure of larger periodic growth bands which collectively form what scientists are currently referring to as a hierarchical composite structure. (The term hierarchy simply implies that there are a range of structural features which exist over a wide range of length scales).

Future developments reside in the synthesis of bio-inspired materials through processing methods and strategies that are characteristic of biological systems. These involve nanoscale self-assembly of the components and the development of hierarchical structures.

Chemometrics

Chemometrics is the science of extracting information from chemical systems by data-driven means. Chemometrics is inherently interdisciplinary, using methods frequently employed in core data-analytic disciplines such as multivariate statistics, applied mathematics, and computer science, in order to address problems in chemistry, biochemistry, medicine, biology and chemical engineering. In this way, it mirrors other interdisciplinary fields, such as psychometrics and econometrics.

Introduction

Chemometrics is applied to solve both descriptive and predictive problems in experimental natural sciences, especially in chemistry. In descriptive applications, properties of chemical systems are modeled with the intent of learning the underlying relationships and structure of the system (i.e., model understanding and identification). In predictive applications, properties of chemical systems are modeled with the intent of predicting new properties or behavior of interest. In both cases, the datasets can be small but are often very large and highly complex, involving hundreds to thousands of variables, and hundreds to thousands of cases or observations.

Chemometric techniques are particularly heavily used in analytical chemistry and metabolomics, and the development of improved chemometric methods of analysis also continues to advance the state of the art in analytical instrumentation and methodology. It is an application-driven discipline, and thus while the standard chemometric methodologies are very widely used industrially, academic groups are dedicated to the continued development of chemometric theory, method and application development.

Origins

Although one could argue that even the earliest analytical experiments in chemistry involved a form of chemometrics, the field is generally recognized to have emerged in the 1970s as computers became increasingly exploited for scientific investigation. The term 'chemometrics' was coined by Svante Wold in a grant application 1971, and the International Chemometrics Society was formed shortly thereafter by Svante Wold and Bruce Kowalski, two pioneers in the field. Wold was a professor of organic chemistry at Umeå University, Sweden, and Kowalski was a professor of analytical chemistry at University of Washington, Seattle.

Many early applications involved multivariate classification, numerous quantitative predictive applications followed, and by the late 1970s and early 1980s a wide variety of data- and computer-driven chemical analyses were occurring.

Multivariate analysis was a critical facet even in the earliest applications of chemometrics. The data resulting from infrared and UV/visible spectroscopy are often easily numbering in the thousands of measurements per sample. Mass spectrometry, nuclear magnetic resonance, atomic emission/absorption and chromatography experiments are also all by nature highly multivariate. The structure of these data was found to be conducive to using techniques such as principal components analysis (PCA), and partial least-squares (PLS). This is primarily because, while the datasets may be highly multivariate there is strong and often linear low-rank structure present. PCA and PLS have been shown over time very effective at empirically modeling the more chemically interesting low-rank structure, exploiting the interrelationships or 'latent variables' in the data, and providing alternative compact coordinate systems for further numerical analysis such as regression, clustering, and pattern recognition. Partial least squares in particular was heavily used in chemometric applications for many years before it began to find regular use in other fields.

Through the 1980s three dedicated journals appeared in the field: Journal of Chemometrics, Chemometrics and Intelligent Laboratory Systems, and Journal of Chemical Information and Modeling. These journals continue to cover both fundamental and methodological research in chemometrics. At present, most routine applications of existing chemometric methods are commonly published in application-oriented journals (e.g., Applied Spectroscopy, Analytical Chemistry, Anal. Chim. Acta., Talanta). Several important books/monographs on chemometrics were also first published in the 1980s, including the first edition of Malinowski's Factor Analysis in Chemistry, Sharaf, Illman and Kowalski's Chemometrics, Massart et al. Chemometrics: a textbook, and Multivariate Calibration by Martens and Naes.

Some large chemometric application areas have gone on to represent new domains, such as molecular modeling and QSAR, cheminformatics, the '-omics' fields of genomics, proteomics, metabonomics and metabolomics, process modeling and process analytical technology.

An account of the early history of chemometrics was published as a series of interviews by Geladi and Esbensen.

Techniques

Multivariate Calibration

Many chemical problems and applications of chemometrics involve calibration. The objective is to develop models which can be used to predict properties of interest based on measured properties of the chemical system, such as pressure, flow, temperature, infrared, Raman, NMR spectra and mass spectra. Examples include the development of multivariate models relating 1) multi-wavelength spectral response to analyte concentration, 2) molecular descriptors to biological activity, 3) multivariate process conditions/states to final product attributes. The process requires a calibration or training data set, which includes reference values for the properties of interest for prediction, and the measured attributes believed to correspond to these properties. For case 1), for example, one can assemble data from a number of samples, including concentrations for an analyte of interest for each sample (the reference) and the corresponding infrared spectrum of that sample. Multivariate calibration techniques such as partial-least squares regression, or principal component regression (and near countless other methods) are then used to construct a mathematical model that relates the multivariate response (spectrum) to the concentration of the analyte of interest, and such a model can be used to efficiently predict the concentrations of new samples.

Techniques in multivariate calibration are often broadly categorized as classical or inverse methods. The principal difference between these approaches is that in classical calibration the models are solved such that they are optimal in describing the measured analytical responses (e.g., spectra) and can therefore be considered optimal descriptors, whereas in inverse methods the models are solved to be optimal in predicting the properties of interest (e.g., concentrations, optimal predictors). Inverse methods usually require less physical knowledge of the chemical system, and at least in theory provide superior predictions in the mean-squared error sense, and hence inverse approaches tend to be more frequently applied in contemporary multivariate calibration.

The main advantages of the use of multivariate calibration techniques is that fast, cheap, or non-destructive analytical measurements (such as optical spectroscopy) can be used to estimate sample properties which would otherwise require time-consuming, expensive or destructive testing (such as LC-MS). Equally important is that multivariate calibration allows for accurate quantitative analysis in the presence of heavy interference by other analytes. The selectivity of the analytical method is provided as much by the mathematical calibration, as the analytical measurement modalities. For example near-infrared spectra, which are extremely broad and non-selective compared to other analytical techniques (such as infrared or Raman spectra), can often be used successfully in conjunction with carefully developed multivariate calibration methods to predict concentrations of analytes in very complex matrices.

Classification, Pattern Recognition, Clustering

Supervised multivariate classification techniques are closely related to multivariate calibration techniques in that a calibration or training set is used to develop a mathematical model capable of classifying future samples. The techniques employed in chemometrics are similar to those used in other fields – multivariate discriminant analysis, logistic regression, neural networks, regression/classification trees. The use of rank reduction techniques in conjunction with these conventional classification methods is routine in chemometrics, for example discriminant analysis on principal components or partial least squares scores.

Unsupervised classification (also termed cluster analysis) is also commonly used to discover patterns in complex data sets, and again many of the core techniques used in chemometrics are common to other fields such as machine learning and statistical learning.

Multivariate Curve Resolution

In chemometric parlance, multivariate curve resolution seeks to deconstruct data sets with limited or absent reference information and system knowledge. Some of the earliest work on these techniques was done by Lawton and Sylvestre in the early 1970s. These approaches are also called self-modeling mixture analysis, blind source/signal separation, and spectral unmixing. For example, from a data set comprising fluorescence spectra from a series of samples each containing multiple fluorophores, multivariate curve resolution methods can be used to extract the fluorescence spectra of the individual fluorophores, along with their relative concentrations in each of the samples, essentially unmixing the total fluorescence spectrum into the contributions from the individual components. The problem is usually ill-determined due to rotational ambiguity (many possible solutions can equivalently represent the measured data), so the application of additional constraints is common, such as non-negatively, unmodality, or known interrelationships between the individual components (e.g., kinetic or mass-balance constraints).

Other Techniques

Experimental design remains a core area of study in chemometrics and several monographs are specifically devoted to experimental design in chemical applications. Sound principles of experimental design have been widely adopted within the chemometrics community, although many complex experiments are purely observational, and there can be little control over the properties and interrelationships of the samples and sample properties.

Signal processing is also a critical component of almost all chemometric applications, particularly the use of signal pretreatments to condition data prior to calibration or classification. The techniques employed commonly in chemometrics are often closely related to those used in related fields.

Performance characterization, and figures of merit Like most arenas in the physical sciences, chemometrics is quantitatively oriented, so considerable emphasis is placed on performance characterization, model selection, verification & validation, and figures of merit. The performance of quantitative models is usually specified by root mean squared error in predicting the attribute of interest, and the performance of classifiers as a true-positive rate/false-positive rate pairs (or a full ROC curve). A recent report by Olivieri et al. provides a comprehensive overview of figures of merit and uncertainty estimation in multivariate calibration, including multivariate definitions of selectivity, sensitivity, SNR and prediction interval estimation. Chemometric model selection usually involves the use of tools such as resampling (including bootstrap, permutation, cross-validation).

Multivariate statistical process control (MSPC), modeling and optimization accounts for a substantial amount of historical chemometric development. Spectroscopy has been used successfully for online monitoring of manufacturing processes for 30–40 years, and this process data is highly amenable to chemometric modeling. Specifically in terms of MSPC, multiway modeling of batch and continuous processes is increasingly common in industry and remains an active area of research in chemometrics and chemical engineering. Process analytical chemistry as it was originally termed, or the newer term process analytical technology continues to draw heavily on chemometric methods and MSPC.

Multiway methods are heavily used in chemometric applications. These are higher-order extensions of more widely used methods. For example, while the analysis of a table (matrix, or second-order array) of data is routine in several fields, multiway methods are applied to data sets that involve 3rd, 4th, or higher-orders. Data of this type is very common in chemistry, for example a liquid-chromatography / mass spectrometry (LC-MS) system generates a large matrix of data (elution time versus m/z) for each sample analyzed. The data across multiple samples thus comprises a data cube. Batch process modeling involves data sets that have time vs. process variables vs. batch number. The multiway mathematical methods applied to these sorts of problems include PARAFAC, trilinear decomposition, and multiway PLS and PCA.

Process Engineering

Process engineering focuses on the design, operation, control, and optimization of chemical, physical, and biological processes. Process engineering encompasses a vast range of industries, such as chemical, petrochemical, agriculture, mineral processing, advanced material, food, pharmaceutical, software development and biotechnological industries.

Process engineering involves translating the needs of the customer into (typically) production facilities that convert "raw materials" into value-added components that are transported to the next stage of the supply chain, typically "packaging engineering", but some larger volume processes such as petroleum refining tend to transfer the products into transportation (trucks or rail) that are then directed to distributors or bulk outlets. Prior to construction, the design work of process engineering begins with a "block diagram" showing raw materials and the transformations/unit operations desired. The design work then progresses to a Process flow diagram (PFD) where material flow paths, storage equipment (such as tanks and silos), transformations/Unit Operations (such as distillation columns, receiver/head tanks, mixing, separations, pumping, etc.) and flowrates are specified, as well as a list of all pipes and conveyors and their contents, material properties such as density, viscosity, particle size distribution, flow rates, pressures, temperatures, and materials of construction for the piping and unit operations. The process flow diagram is then used to develop a Piping and instrumentation diagram (P&ID) which includes pipe and conveyor sizing information to address the desired flowrates, process controls (such as tank level indications, material flow meters, weighing devices, motor speed controls, temperature and pressure indicators/controllers, etc.). The P&ID is then used as a basis of design for developing the "system operation guide" or "functional design specification" which outlines the operation of the process. From the "P&ID", a proposed layout (general arrangement) of the process can be shown from an overhead view (plot plan) and a side view (elevation), and other engineering disciplines are involved such as civil engineers for site work (earth moving), foundation design, concrete slab design work, structural steel to support the equipment, etc.). All previous work is directed toward defining the scope of the project, then developing a cost estimate to get the design installed, and a schedule to communicate the timing needs for engineering, procurement, fabrication, installation, commissioning, startup, and ongoing production of the process. Depending on the needed accuracy of the cost estimate and schedule that is required, several iterations of designs are generally provided to customers or stakeholders who feedback their requirements and the process engineer incorporates these additional instructions and wants (scope revisions) into the overall design and additional cost estimates and Schedules are developed for funding approval. Following funding approval, the project is executed via project management.

The application of systematic computer-based methods to process engineering is process systems engineering.

Significant Accomplishments

Several accomplishments have been made in Process Systems Engineering:

- Process design: synthesis of energy recovery networks, synthesis of distillation systems (azeotropic), synthesis of reactor networks, hierarchical decomposition flowsheets, superstructure optimization, design multiproduct batch plants. Design of the production reactors for the production of plutonium, design of nuclear submarines.

- Process control: model predictive control, controllability measures, robust control, nonlinear control, statistical process control, process monitoring, thermodynamics-based control

- Process operations: scheduling process networks, multiperiod planning and optimization, data reconciliation, real-time optimization, flexibility measures, fault diagnosis

- Supporting tools: sequential modular simulation, equation based process simulation, AI/expert systems, large-scale nonlinear programming (NLP), optimization of differential algebraic equations (DAEs), mixed-integer nonlinear programming (MINLP), global optimization

History of Process Systems Engineering

Process systems engineering (PSE) is a relatively young area in chemical engineering. The first time that this term was used was in a Special Volume of the AIChE Symposium Series in 1961. However, it was not until 1982 when the first international symposium on this topic took place in Kyoto, Japan, that the term PSE started to become widely accepted.

The first textbook in the area was "Strategy of Process Engineering" by Dale F. Rudd and Charles C. Watson, Wiley, 1968. The Computing and Systems Technology (CAST) Division, Area 10 of AIChE, was founded in 1977 and currently has about 1200 members. CAST has four sections: Process Design, Process Control, Process Operations, and Applied Mathematics.

The first journal devoted to PSE was "Computers and Chemical Engineering," which appeared in 1977. The Foundations of Computer-Aided Process Design (FOCAPD) conference in 1980 in Henniker was one of the first meetings in a series on that topic in the PSE area. It is now accompanied by the successful series on Control (CPC), Operations (FOCAPO), and the world-wide series entitled Process Systems Engineering. The CACHE Corporation (Computer Aids for Chemical Engineering), which organizes these conferences, was initially launched by academics in 1970, motivated by the introduction of process simulation in the chemical engineering curriculum.

Roger W.H. Sargent from Imperial College was one of the pioneers in the area. PSE is an active area of research in many other countries, particularly in the United Kingdom, Germany, Japan, Korea, and China.

References

- Singh , R Paul; Dennis R. Heldman (2013). Introduction to Food Engineering (5th ed.). Academic Press. p. 1. ISBN 0123985307.

- von Hippel; A. R. (1954). "Ceramics". Dielectric Materials and Applications. Technology Press (M.I.T.) and John Wiley & Sons. ISBN 1-58053-123-7.

- Harris, D.C., "Materials for Infrared Windows and Domes: Properties and Performance", SPIE PRESS Monograph, Vol. PM70 (Int. Society of Optical Engineers, Bellingham WA, 2009) ISBN 978-0-8194-5978-7

- Brinker, C.J.; Scherer, G.W. (1990). Sol-Gel Science: The Physics and Chemistry of Sol-Gel Processing. Academic Press. ISBN 0-12-134970-5.

- Malinowski, E. R.; Howery, D. G. (1980). Factor Analysis in Chemistry. New York: Wiley. ISBN 0471058815.

- Sharaf, M. A.; Illman, D. L.; Kowalski, B. R., eds. (1986). Chemometrics. New York: Wiley. ISBN 0471831069.

- Massart, D. L.; Vandeginste, B. G. M.; Deming, S. M.; Michotte, Y.; Kaufman, L. (1988). Chemometrics: a textbook. Amsterdam: Elsevier. ISBN 0444426604.

- Franke, J. (2002). "Inverse Least Squares and Classical Least Squares Methods for Quantitative Vibrational Spectroscopy". Handbook of Vibrational Spectroscopy. New York: Wiley. doi:10.1002/0470027320.s4603. ISBN 0471988472.

- Hunter, W. G. (1984). "Statistics and chemistry, and the linear calibration problem". In Kowalski, B. R. Chemometrics: mathematics and statistics in chemistry. Boston: Riedel. ISBN 9027718466.

- Bruns, R. E.; Scarminio, I. S.; de Barros Neto, B. (2006). Statistical design – chemometrics. Amsterdam: Elsevier. ISBN 044452181X.

Softwares Related to Chemical Engineering

Chemical workbench is a software that aims at the scale of homogeneous gas-phase and heterogeneous processes whereas COCO simulator is a free-of-charge and sequential simulation process modeling environment. In order to completely understand chemical engineering, it is necessary to understand the softwares related to it. The following text elucidates the varied softwares used in chemical engineering.

Chemical WorkBench

Chemical WorkBench is a proprietary simulation software tool aimed at the reactor scale kinetic modeling of homogeneous gas-phase and heterogeneous processes and kinetic mechanism development. It can be effectively used for the modeling, optimization, and design of a wide range of industrially and environmentally important chemistry-loaded processes. Chemical WorkBench is a modeling environment based on advanced scientific approaches, complementary databases, and accurate solution methods. Chemical WorkBench is developed and distributed by Kintech Lab.

Chemical Workbench Models

Chemical WorkBench has an extensive library of physicochemical models:

- Thermodynamic Models

- Gas-Phase Kinetic Models

- Flame model

- Heterogeneous Kinetic Models

- Non-Equilibrium Plasma Models

- Detonation and Aerodynamic Models

- Membrane Separation Models

- Mechanism Analysis and Reduction

Fields of Application

Chemical WorkBench can be used by researchers and engineers working in the following fields:

- General chemical kinetics and thermodynamics

- Kinetic mechanisms development

- Thin films growth for microelectronics

- Nanotechnology

- Catalysis and chemical engineering

- Combustion, detonation and pollution control

- Waste treatment and recovering

- Plasma light sources and plasma chemistry

- High-temperature chemistry

- Education

- Combustion and detonation, clean power-generation technologies, safety analysis, CVD, heterogeneous and catalytic reactions and processes, and processes in non-equilibrium plasmas are the main areas of interest.

COCO Simulator

The COCO Simulator is a free-of-charge, non-commercial, graphical, modular and CAPE-OPEN compliant, steady-state, sequential simulation process modeling environment. It was originally intended as a test environment for CAPE-OPEN modeling tools but now provides free chemical process simulation for students. It is an open flowsheet modeling environment allowing anyone to add new unit operations or thermodynamics packages.

The COCO Simulator uses a graphical representation, the Process Flow Diagram (PFD), for defining the process to be simulated. Clicking on a unit operation with the mouse allows the user to edit the unit operation parameters it defines via the CAPE-OPEN standard or to open the unit operation's own user interface, when available. This interoperability of process modeling software was enabled by the advent of the CAPE-OPEN standard. COCO thermodynamic library "TEA" and its chemical compound data bank are based on ChemSep LITE, a free equilibrium column simulator for distillation columns and liquid-liquid extractors. COCO's thermodynamic library exports more than 100 property calculation methods with their analytical or numerical derivatives.

COCO includes a LITE version of COSMOtherm, an activity coefficient model based on Ab initio quantum chemistry methods. The simulator entails a set of unit-operations such as stream splitters/mixers, heat-exchangers, compressors, pumps and reactors. COCO features a reaction numerics package to power its simple conversion, equilibrium, CSTR, Gibbs minimization and plug flow reactor models.

DWSIM

DWSIM is an open-source CAPE-OPEN compliant chemical process simulator for Windows and Linux. DWSIM is built on top of the Microsoft .NET and Mono Platforms and features a Graphical User Interface (GUI), advanced thermodynamics calculations, reactions support and petroleum characterization / hypothetical component generation tools.

DWSIM is able to simulate steady-state, vapor–liquid, vapor–liquid-liquid, solid–liquid and aqueous electrolyte equilibrium processes with the following Thermodynamic Models and Unit Operations:

- Thermodynamic models: PC-SAFT, FPROPS, CoolProp, Peng–Robinson equation of state, Peng–Robinson-Strÿjek-Vera (PRSV2), Soave–Redlich–Kwong, Lee-Kesler, Lee-Kesler-Plöcker, UNIFAC(-LL), Modified UNIFAC (Dortmund), Modified UNIFAC (NIST), UNIQUAC, NRTL, COSMO-SAC, Chao-Seader, Grayson-Streed, Extended UNIQUAC, Raoult's Law, IAPWS-IF97 Steam Tables, IAPWS-08 Seawater, Black-Oil and Sour Water;

- Unit operations: CAPE-OPEN Socket, Spreadsheet, Custom (IronPython Script), Mixer, Splitter, Separator, Pump, Compressor, Expander, Heater, Cooler, Valve, Pipe Segment, Shortcut Column, Heat exchanger, Reactors (Conversion, PFR, CSTR, Equilibrium and Gibbs), Distillation column, Simple, Refluxed and Reboiled Absorbers, Component Separator, Solids Separator, Continuous Cake Filter and Orifice plate;

- Utilities: Binary Data Regression, Phase Envelope, Natural Gas Hydrates, Pure Component Properties, True Critical Point, PSV Sizing, Vessel Sizing, Spreadsheet and Petroleum Cold Flow Properties;

- Tools: Hypothetical Component Generator, Bulk C7+/Distillation Curves Petroleum Characterization, Petroleum Assay Manager, Reactions Manager and Compound Creator;

- Process Analysis and Optimization: Sensitivity Analysis Utility, Multivariate Optimizer with bound constraints;

- Extras: Support for Runtime Scripts, Plugins and CAPE-OPEN Flowsheet Monitoring Objects.

Android and iOS Versions

DWSIM is also available on Android and iOS mobile operating systems, where it is free to download. On these platforms, DWSIM includes a basic set of features while more advanced modules can be unlocked through in-app purchases.

Standalone Thermodynamics Library

DWSIM's Property and Equilibrium calculation routines are also available as a standalone, 100% managed .NET Dynamic Link Library (DLL). It can be linked against free and proprietary applications (LGPL v3 license).

ProMax

ProMax is a chemical process simulator for process troubleshooting and design, developed and sold by Bryan Research and Engineering, Inc. Initially released in late 2005, ProMax is a continuance of two previous process simulators, PROSIM and TSWEET. ProMax is considered the industry standard for designing amine gas treating and glycol dehydration units.

Program History

In 1974 Bryan Research and Engineering (BR&E) began developing simulation software for sulfur recovery units with a command-line interface. In 1976 this program was released under the name SULFUR. Amine sweetening, for which BR&E is most well known, was added in 1978 and the simulation package was renamed TSWEET. A second product, DEHY, was released in 1980 for modeling glycol dehydration units. Natural gas processing was added to the DEHY program in 1983 and the package was renamed PROSIM. In 1988 BR&E introduced a graphical user interface to both programs; a novelty for chemical process simulators at the time.

TSWEET and PROSIM were both MS-DOS based programs and were both incorporated into ProMax. ProMax is a late generation Windows application which uses Microsoft Visio as the graphical user interface. Other capabilities were included in ProMax besides those already available in TSWEET and PROSIM enabling it to model almost any process in the oil and gas industry.

Company History

Bryan Research & Engineering, Inc. (BR&E) is a privately owned provider of software and engineering solutions to the oil, gas, refining and chemical industries. Since the company's inception in 1974, BR&E has combined research and development in process simulation to provide clients with simulation tools.

References

- W.D. Seider; J.D. Seader; D.R. Lewin (1999). Process Design Principles. Wiley. ISBN 0-471-24312-4.

- J.M. Douglas (1988). Conceptual Design of Chemical Processes. McGraw-Hill. ISBN 0-07-017762-7.

- W.L. McCabe; J.C. Smith; P. Harriot (1993). Unit Operations of Chemical Engineering (5th ed.). McGraw-Hill. ISBN 0-07-044844-2.

- Perry, Robert H.; Green, Don W. (1997). Perry's Chemical Engineers' Handbook (7th ed.). McGraw-Hill. ISBN 0-07-049841-5. p. 13-53

Evolution of Chemical Engineering

Chemical engineering has evolved over the years. The Industrial Revolution was a major landmark in the mass manufacture of chemicals. Chemical engineering in contemporary times is used in numerous fields, and is studied in a number of universities. This chapter has been carefully written to provide an easy understanding of the development of chemical engineering.

Chemical engineering as a discipline that was developed out of those practising "industrial chemistry" in the late 19th century. Before the Industrial Revolution (18th century), industrial chemicals and other consumer products such as soap were mainly produced through batch processing. Batch processing is labour-intensive and individuals mix predetermined amounts of ingredients in a vessel, heat, cool or pressurize the mixture for a predetermined length of time. The product may then be isolated, purified and tested to achieve a saleable product. Batch processes are still performed today on higher value products, such as pharmaceutical intermediates, speciality and formulated products such as perfumes and paints, or in food manufacture such as pure maple syrups, where a profit can still be made despite batch methods being slower and inefficient in terms of labour and equipment usage. Due to the application of Chemical Engineering techniques during manufacturing process development, larger volume chemicals are now produced through a continuous "assembly line" chemical processes. The Industrial Revolution was when a shift from batch to more continuous processing began to occur. Today commodity chemicals and petrochemicals are predominantly made using continuous manufacturing processes whereas speciality chemicals, fine chemicals and pharmaceuticals are made using batch processes.

Origin

The Industrial Revolution led to an unprecedented escalation in demand, both with regard to quantity and quality, for bulk chemicals such as soda ash. This meant two things: one, the size of the activity and the efficiency of operation had to be enlarged, and two, serious alternatives to batch processing, such as continuous operation, had to be examined.

The First Chemical Engineer

Industrial chemistry was being practised in the mid 1800s, but it was not until the 1880s that the engineering elements required to control chemical processes were being recognized as a distinct professional activity. Chemical engineering was first established as a profession in the United Kingdom when the first chemical engineering course was given at the University of Manchester in 1887 by George E. Davis in the form of twelve lectures covering various aspects of industrial chemical practice. As a consequence George E. Davis is regarded as the world's first chemical engineer. Today, chemical engineering is a highly regarded profession. Chemical engineers with experience can become licensed Professional Engineers in the United States, aided by the National Society of Professional Engineers, or gain "Chartered" chemical-engineer status through the UK-based Institution of Chemical Engineers.

Professional Associations

In 1880, the first attempt was made to form a Society of Chemical Engineers in London. This eventually resulted in the formation of the Society of Chemical Industry in 1881. The American Institute of Chemical Engineers (AIChE) was founded in 1908, and the UK Institution of Chemical Engineers (IChemE) in 1922. These both now have substantial international membership. Some other countries now have chemical engineering societies or sections within chemical or engineering societies, but the AIChE and IChemE remain the major ones in numbers and international spread: they are both open to suitably qualified professionals or students of chemical engineering anywhere in the world.

Definitions

For the other established branches of engineering, there were ready associations in the mind of the common man: Mechanical Engineering meant machines, Electrical Engineering meant circuitry, and Civil Engineering meant structures. So chemical engineering can be symbolised as chemicals production.

Unit Operation

The answer, provided by Arthur D. Little to the president of MIT, was to emphasize the approach chemical engineers took to the design and analysis of processes rather than a process or a product. The concept of Unit operations was developed to emphasize the underlying unity among seemingly different operations. For example, the principles are the same whether one is concerned about separating alcohol from water in a fermenter, or separating gasoline from diesel in a refinery, as long as the basis of separation is generation of a vapor of a different composition from the liquid. Therefore, such separation processes can be studied together as a unit operation (in this case called distillation). The concept has stood the profession in good stead in its phase of growth, and has even been used to understand the way the human body functions.

Unit Processes

In the early part of the last century, a parallel concept called Unit Processes was used to classify reactive processes. Thus oxidations, reductions, alkylations, etc. formed separate unit processes and were studied as such. This was natural considering the close affinity of chemical engineering to industrial chemistry at its inception. Gradually however, the subject of chemical reaction engineering has largely replaced the unit process concept. This subject looks at the entire body of chemical reactions as having a personality of its own, independent of the particular chemical species or chemical bonds involved. The latter does contribute to this personality in no small measure, but to design and operate chemical reactors, a knowledge of characteristics such as rate behaviour, thermodynamics, single or multiphase nature, etc. are more important. The emergence of chemical reaction engineering as a discipline truly signaled the severance of the umbilical cord connecting chemical engineering to industrial chemistry, and served to cement the truly unique character of this discipline.

Permissions

All chapters in this book are published with permission under the Creative Commons Attribution Share Alike License or equivalent. Every chapter published in this book has been scrutinized by our experts. Their significance has been extensively debated. The topics covered herein carry significant information for a comprehensive understanding. They may even be implemented as practical applications or may be referred to as a beginning point for further studies.

We would like to thank the editorial team for lending their expertise to make the book truly unique. They have played a crucial role in the development of this book. Without their invaluable contributions this book wouldn't have been possible. They have made vital efforts to compile up to date information on the varied aspects of this subject to make this book a valuable addition to the collection of many professionals and students.

This book was conceptualized with the vision of imparting up-to-date and integrated information in this field. To ensure the same, a matchless editorial board was set up. Every individual on the board went through rigorous rounds of assessment to prove their worth. After which they invested a large part of their time researching and compiling the most relevant data for our readers.

The editorial board has been involved in producing this book since its inception. They have spent rigorous hours researching and exploring the diverse topics which have resulted in the successful publishing of this book. They have passed on their knowledge of decades through this book. To expedite this challenging task, the publisher supported the team at every step. A small team of assistant editors was also appointed to further simplify the editing procedure and attain best results for the readers.

Apart from the editorial board, the designing team has also invested a significant amount of their time in understanding the subject and creating the most relevant covers. They scrutinized every image to scout for the most suitable representation of the subject and create an appropriate cover for the book.

The publishing team has been an ardent support to the editorial, designing and production team. Their endless efforts to recruit the best for this project, has resulted in the accomplishment of this book. They are a veteran in the field of academics and their pool of knowledge is as vast as their experience in printing. Their expertise and guidance has proved useful at every step. Their uncompromising quality standards have made this book an exceptional effort. Their encouragement from time to time has been an inspiration for everyone.

The publisher and the editorial board hope that this book will prove to be a valuable piece of knowledge for students, practitioners and scholars across the globe.

Index

A

Acid-base Reactions, 64, 70-71, 74
Affinity Chromatography, 136, 189
Air-sensitive Vacuum Distillation, 163
Allotropes, 11, 16, 20-21, 33
Asymmetric Flow Fff (af4), 185
Atomic Mass, 19, 26
Atomic Number, 15-20, 22-24, 27-28, 32, 34-35

B

Batch Distillation, 157-160, 168, 170, 203
Biochemical Engineering, 3, 40

C

Catalytic Reactor, 199-201
Ceramic Engineering, 234, 253-254, 256
Chemical Compounds, 8, 10-12, 14, 16, 42, 100, 139, 149, 253
Chemical Elements, 1, 8, 10-11, 13-16, 20, 22-29, 31-34, 70
Chemical Equilibrium, 64, 84, 103-104, 224
Chemical Kinetics, 190, 196, 198, 201, 216, 220, 224-225, 233, 279
Chemical Plant, 4-5, 42, 201-207, 209
Chemical Process, 6, 13, 40, 42-43, 49, 121-122, 188, 202, 205, 208, 210, 217, 228-230, 279-281
Chemical Process Modeling, 40, 42, 122
Chemical Reaction, 4, 9, 11-13, 15, 25, 42, 44, 51, 60-61, 63, 71, 73, 81, 83-84, 88, 92, 100, 103, 119, 154, 166, 170, 196, 199, 201, 212-213, 215, 217, 220-225, 285
Chemical Reaction Engineering, 4, 119, 196, 285
Chemical Reactor, 7, 190, 196-197, 200-201, 225, 229
Chemical Substance, 8-16, 126, 212-213
Chemical Thermodynamics, 4, 210-213, 216, 232
Chemical Workbench, 278-279
Chemically Pure, 20
Chemicals Versus Chemical Substances, 13
Cheminformatics, 7, 234, 272
Chemometrics, 270-274, 277
Chromatographic Processes, 189

Chromatography, 114, 126-127, 129-141, 181, 188-189, 271, 275
Clustering of Commodity Chemical Plants, 209
Coco Simulator, 278-279
Column Chromatography, 133
Construction, 4-6, 40, 59, 120-121, 205, 210, 220, 230-232, 243, 249, 275
Continuous Distillation, 157, 159-160, 203
Cooling Crystallization, 149-150
Crystal Structures, 20, 22, 143
Crystallization, 5, 7, 43, 65, 118-119, 124, 127, 142-152, 164, 166, 173, 258, 269

D

Decomposition, 67-68, 82, 96, 98, 140, 163, 167, 225, 275-276
Dehydrogenation, 43, 59, 98
Densities, 20-21, 23, 263
Displacement Chromatography, 134
Distillation, 1, 7, 43, 119-120, 124, 126-128, 153-173, 176, 179, 186-189, 200, 203-204, 219, 275-276, 279-280, 284
Double Replacement, 67-68
Dtb Crystallizer, 152-153
Dwsim, 217, 280-281

E

Electrochemical Engineering, 2, 4, 6, 8, 10, 12, 14, 16, 18, 20, 22, 24, 26, 28, 30, 32, 34, 36, 38, 40-41
Elementary Reactions, 60, 63-64, 221
Evaporative Crystallization, 149, 151
Evaporative Crystallizers, 152
Expanded Bed Adsorption Chromatographic Separation, 137

F

Field Flow Fractionation, 127, 181-182
Flow Geometries, 176
Food Engineering, 6, 234-235, 277
Fractional Distillation, 128, 155, 161, 168

G

Gas Chromatography, 129-130, 135, 140

H
Hollow Fiber Flow Fff, 185
Hydrodynamic Model, 174-175
Hydrogenation, 43, 46, 82, 85, 90-91

I
Idealized Distillation Model, 157
Ion Exchange Chromatography, 136-137
Isotopes, 11, 15, 17-20, 22, 26, 33, 181, 225
Isotopic Mass, 19
Isotopically Pure, 20

K
Kinetics, 41, 66, 94, 190, 196, 198, 201, 216, 219-225, 232-233, 279

L
Laboratory Scale Distillation, 157, 160
Liquid Chromatography, 130, 132, 135-136, 138-141

M
Membrane Separation Processes, 173, 179
Membrane Shapes, 176
Membrane Technology, 173-174

N
Nucleation, 142-147, 258

O
Oxidation, 5, 43-55, 63, 68-70, 74, 77, 81, 85, 87, 90-91, 96, 98-99

P
Paper Chromatography, 130, 133-134
Pfr (plug Flow Reactor), 199
Pharmaceutical Engineering, 237-238
Photochemical Reactions, 72
Planar Chromatography, 133
Plant Design, 4, 204-207
Plastics Engineering, 6, 234, 236
Primary Nucleation, 146
Process Design, 3, 5-6, 41-42, 120, 122-124, 189, 228, 230, 239, 276-277, 282
Process Engineering, 6, 225, 234, 275-277
Process Flow Diagram, 5, 122, 218, 225-227, 275, 279

Process Integration, 42, 122-124
Process Miniaturization, 6, 228-232
Process Simulation, 190, 217-220, 230, 276-277, 279, 281
Promax, 281
Pyrolysis Gas Chromatography, 140

R
Raw Material, 7-8, 97, 162, 196, 202
Reactions At The Solid|gas Interface, 72
Reduction, 43-47, 49-50, 52-55, 61, 65, 68-69, 74, 87, 90-91, 94-95, 165, 232, 235-236, 256, 273, 278
Reversed-phase Chromatography, 138

S
Secondary Nucleation, 147
Semi-batch Reactor, 200
Separation Process, 126-127, 129, 131, 133, 135, 137, 139, 141, 143, 145, 147, 149, 151, 153-155, 157, 159-161, 163, 165, 167, 169, 171, 173, 175-179, 181, 183, 185-187, 189, 200, 232
Simple Distillation, 161
Single Replacement, 67-68
Size-exclusion Chromatography, 137
Solid-state Reactions, 71
Split Flow Thin-cell Fractionation (splitt), 185
States of Matter, 20-21
Steam Distillation, 162, 168
Stefan Tube, 190, 195
Substances Versus Mixtures, 12
Supercritical Fluid Chromatography, 136
Synthesis, 7, 10, 14, 33, 40-41, 53, 61-64, 67-68, 73, 80, 85, 87, 89-90, 92, 96-101, 114, 122, 125, 154, 168, 217, 245, 254, 261, 264, 270, 276

T
The Periodic Table, 16, 18, 21-23, 30, 35
Theoretical Plate, 161, 171, 186, 188-189
Thermal Fff, 185
Thermodynamic View, 144
Thermodynamics, 4, 7, 65, 73, 84-85, 105-106, 145, 190-191, 194, 210-213, 216, 221, 223-224, 232, 276, 279-281, 285
Thin Layer Chromatography (tlc), 134
Tissue Engineering, 234, 239-241, 243-244, 246, 248-252, 257

Transport Phenomena, 3, 5, 7, 190-192, 232

U

Unit Operation, 5, 42, 118-119, 154, 279, 284

V

Vacuum Distillation, 162-165, 171-172

Z

Zone Distillation, 164

* 9 7 8 1 6 3 5 4 9 0 6 3 3 *